Introduction to Self-Driving Vehicle Technology

T0144258

Chapman & Hall/CRC
Artificial Intelligence and Robotics Series

Series Editor: Roman Yampolskiy

Intelligent Autonomy of UAVs
Advanced Missions and Future Use
Yasmina Bestaoui Sebbane

Artificial Intelligence
With an Introduction to Machine Learning, Second Edition
Richard E. Neapolitan, Xia Jiang

Artificial Intelligence and the Two Singularities
Calum Chace

Behavior Trees in Robotics and Al
An Introduction
Michele Collendanchise, Petter Ögren

Artificial Intelligence Safety and Security
Roman V. Yampolskiy

Artificial Intelligence for Autonomous Networks
Mazin Gilbert

Virtual Humans
David Burden, Maggi Savin-Baden

Deep Neural Networks: WASD Neuronet Models, Algorithms, and Applications
Yunong Zhang, Dechao Chen, Chengxu Ye

Introduction to Self-Driving Vehicle Technology
Hanky Sjafrie

For more information about this series please visit:
https://www.crcpress.com/Chapman--HallCRC-Artificial-Intelligence-and-Robotics-Series/book-series/ARTILRO

Introduction to Self-Driving Vehicle Technology

Hanky Sjafrie

CRC Press
Taylor & Francis Group
Boca Raton London New York

CRC Press is an imprint of the
Taylor & Francis Group, an **informa** business

A CHAPMAN & HALL BOOK

Illustration by Kalpana Tarte

CRC Press
Taylor & Francis Group
6000 Broken Sound Parkway NW, Suite 300
Boca Raton, FL 33487-2742

© 2020 by Taylor & Francis Group, LLC
CRC Press is an imprint of Taylor & Francis Group, an Informa business

No claim to original U.S. Government works

Printed on acid-free paper

International Standard Book Number-13: 978-0-367-32126-0 (Hardback)
                                      978-0-367-32125-3 (Paperback)

Library of Congress Cataloging-in-Publication Data

Names: Sjafrie, Hanky, author.
Title: Introduction to self-driving vehicle technology / by Hanky Sjafrie.
Description: Boca Raton, Florida : CRC Press, [2020] |
Series: Chapman & Hall/CRC artificial intelligence and robotics series | Includes
bibliographical references and index. |
Summary: "This book aims to teach the core concepts that make Self-driving vehicles
(SDVs) possible. It is aimed at people who want to get their teeth into self-driving vehicle
technology, by providing genuine technical insights where other books just skim the
surface. The book tackles everything from sensors and perception to functional safety
and cybersecurity. It also passes on some practical know-how and discusses concrete SDV
applications, along with a discussion of where this technology is heading" -- Provided
by publisher.
Identifiers: LCCN 2019026764 (print) | LCCN 2019026765 (ebook) | ISBN 9780367321260
(hardback) | ISBN 9780367321253 (paperback) | ISBN 9780429316777 (ebook)
Subjects: LCSH: Autonomous vehicles -- Technological innovations. |
Automobiles–Automatic control. | Automobile industry and trade -- Technological
innovations.
Classification: LCC TL152.8 .S57 2020 (print) | LCC TL152.8 (ebook) |
DDC 629.2/046--dc23
LC record available at https://lccn.loc.gov/2019026764
LC ebook record available at https://lccn.loc.gov/2019026765

Visit the Taylor & Francis Web site at
http://www.taylorandfrancis.com

and the CRC Press Web site at
http://www.crcpress.com

S. D. G.

Contents

Preface

Self-driving vehicles, or SDVs, are a hot topic currently. However, SDVs are based on complex technology, and it can be frustrating trying to get hold of information on exactly how they work. This book sets out to teach you the core concepts that make SDVs possible. It is aimed at people who want to get their teeth into self-driving vehicle technology, providing genuine technical insights where other books just skim the surface.

If you are a software developer or professional engineer who is eager to pursue a career in this exciting field and wants to learn more about the basics of SDV algorithms, then this book will be a good starting point. Likewise, if you are an academic researcher who is keen to apply your expertise in SDVs and wants to know what it takes to build an SDV prototype, then this book should answer your questions. But this book is also suitable for all technology enthusiasts and journalists who want a clear and readable overview of the technologies that make SDVs possible. It covers all the bases, tackling everything from sensors and perception to functional safety and cybersecurity. It also passes on some practical know-how and discusses concrete SDV applications along with a discussion of where this technology is heading.

Sadly, there is a distinct reluctance to divulge technical details in this emerging industry. SDV companies are unwilling to share details of their systems or the data they collect while testing their vehicles. This data is regarded as a valuable commodity in the race to teach cars to drive themselves. Hopefully, this book will go some way to redress this imbalance in information sharing.

Author

Hanky Sjafrie is the CEO of SGEC, an independent engineering consulting firm specializing in automotive software engineering for Advanced Driver Assistance Systems (ADAS) and Autonomous Driving (AD). Much of his experience was acquired through his deep involvement in these fields while working on various research and development (R&D) projects for car manufacturers and automotive technology suppliers, ranging from sensor technologies (radars, lidars, ultrasonics, etc.) to automotive cybersecurity.

Prior to SGEC, he was actively involved in diverse series R&D projects within the domains of ADAS/AD and infotainment systems at BMW and Audi, as well as at a Silicon Valley-based autonomous driving start-up. Besides working with clients from the automotive industry, he also provides insights into the realm of automotive technology to Siemens, Boston Consulting Group, PricewaterhouseCoopers, and Roland Berger, among others.

Acknowledgment

Writing a technical book on a technology that is still in its infancy is never an easy task. This book has benefited from the labors of many outstanding scientists and countless hard-working industry professionals who pionereed the research and development of self-driving vehicle technology. My sincere appreciation goes to them and to many others who work day-in and day-out to make this technology closer to reality.

I want to express my deepest gratitude to my loving wife, Heidi, and my children, Hugo and Helene. Without their constant encouragement and their patience during many weekends and evenings spent working on the book's material, this book would not have been possible.

My special thanks go to my technical reviewers, Stefan Schlichthaerle, Daniel Schwoerer, Mohamed Kemal, Miguel Julia, Martin Schwoerer, and Reinhard Miller. I am deeply grateful for their efforts and the time taken from their busy professional lives to provide comments and improvement suggestions about the book's material.

I cannot thank enough my senior acquisitions editor, Randi Cohen, without whose invaluable input and continued assistance this book would not be nearly as polished as it is today.

Profound gratitude to my graphics illustrator, Kalpana Tarte, for her tremendous effort in creating concise illustrations throughout the book. Thanks to my technical editors, Jason Theophilus, James Humphrey, and Toby Moncaster, for their valuable contributions to turn the complex technical discussions throughout the book into clear and understandable text. Heartfelt thanks go to Anita Rachmat for her early guidance in book publishing.

Last but not least, I wish to express my sincere gratitude to so many friends and colleagues; I cannot possibly hope to name them all individually. I am so grateful for all their support, guidance, and encouragement over the years.

Chapter 1

Introduction

For centuries, people dreamed of flying to the moon. But it wasn't until the Cold War that the two Space Race rivals, the US and the USSR, finally developed the aerospace capabilities to make this dream a reality.

The concept of self-driving vehicles, or SDVs, also stretches back a long time, first appearing in historical records around 1478. In the six centuries since then, the idea of autonomous vehicles has been regularly dismissed as absurd by some, and cherished as a dream by others. But now the momentum for SDVs is building, following a similar trajectory to that of space travel in the late 1950s. Things that once seemed impossible are coming within reach at astonishing speed.

Mention self-driving vehicles, and most people immediately think of autonomous cars. The idea of a car that drives itself has captured the popular imagination. Huge efforts are being made to lay the groundwork for the era of fully autonomous vehicles and hardly a day goes by without automakers, tech companies, policymakers, insurance brokers, or infrastructure firms unveiling the latest new development.

This book tackles the subject of SDVs from a technical perspective. It explains the big picture of what SDVs are and how they are developed, while offering plenty of in-depth engineering insights along the way.

1.1 Brief history of SDV technology

The concept of a self-driving vehicle may seem like a purely modern invention, but a sketch made by Leonardo da Vinci over 500 years ago suggests otherwise. The drawing shows a self-propelled cart powered by coiled springs and featuring programmable steering based on an arrangement of wooden pegs. In 2004, Paolo Galluzzi, director of the Institute and Museum of the History of Science in Florence, oversaw a project to build a working model based on the design created by da Vinci around 1478. A video shows their painstakingly crafted machine in action [3]. Often cited as the first example of a self-propelled vehicle and programmable machine, the design could also be regarded as the world's first robot vehicle, since it had no driver.

The idea of SDVs resurfaced at the 1939 New York World's Fair in the Futurama exhibit sponsored by the General Motors Corporation. The installation captured the public's imagination with its vision of the world 20 years into the future, showing self-driving cars operating on an automated highway system. Sixteen years later, General Motors expanded on the themes of smart roads and driverless cars in its musical short titled, "Key to the Future." The film of a relentlessly cheerful family enjoying the wonders of autopilot was exhibited at its 1956 Motorama auto show, which was attended by over 2.2 million visitors in various locations around the US [7].

Fast forward another half a century, and the development of autonomous vehicles gained significant momentum with the DARPA Grand Challenge competition held by the US Department of Defense in 2005, and the DARPA Urban Challenge in 2007. In each case, the participating teams were required to build a driverless vehicle, and complete a course within a specified timeframe. This provided a major boost to technologies such as vehicle software and robotics development and marked a tipping point for technological progress in the field of SDVs. Since then, BMW, Audi, Daimler, Google, Tesla, Uber, Baidu, and many other companies have continued to advance autonomous vehicle technology in various ways.

Meanwhile, policymakers in many countries have begun preparing new regulations for the future of autonomous vehicles. From insurance and standards to infrastructure and enabling technologies, the entire automotive ecosystem is playing an active role in embracing the changes that lie ahead.

1.2 What is an SDV?

Describe a vehicle as self-driving, autonomous or driverless, and different people will picture very different things. Some may interpret it as a vehicle that travels completely independently, without requiring a driver at all. Others may imagine a vehicle that travels on its own making autonomous decisions, but that still requires a human driver sitting behind the wheel ready to act in the event of an emergency.

The whole concept of a car driving on its own may seem extraordinarily complex, but the idea behind it is quite simple and lies well within the remit of current technology. Consider, for a moment, the processes involved in driving a vehicle. To begin with, you choose your destination and decide how to get there from your starting point. Then you start driving, keeping an eye on your surroundings at all times. These include static objects such as buildings, trees, road signs, and parked cars, as well as dynamic objects, such as pedestrians, moving cars, and animals. Every now and then, one of these objects may block the road, which will require you to react in one way or another. In the meantime, you, as the driver, are using all the available mechanisms within the vehicle to propel it in the required direction, while obeying the rules of the road.

Described in this way, it seems feasible that some of these steps could be performed without human intervention, and indeed many of the vehicles around us today are automated to a certain degree. Airplanes, trains, and ships all exhibit some degree of autonomy. By applying the computing power we already have at our disposal, and combining it with reliable sensors, intelligent algorithms as well as other components, it is possible to replicate the act of driving.

To provide clarity and a level playing field in the SDV industry, bodies such as the *US Department of Transportation's National Highway Traffic Safety Administration (NHTSA)* and the *German Federal Highway Research Institute (BASt)* have published definitions of SDVs based on their degree of automation. One of the most widely used references is the *J3016 Driving Automation Taxonomy* [4] published by *Society for Automotive Engineers (SAE) International*, which describes six levels of autonomous vehicles ranging from no automation to full automation as summarized in Table 1.1.

It starts with *Level 0*, where the driver performs and monitors all aspects of the driving task and the vehicle's systems only intervene to provide warnings. *Level 1* adds *Driver Assistance* functions, enabling the system to take over either lateral (steering) or longitudinal (acceleration/deceleration) control. The driver still decides when this should

Table 1.1   SAE Levels of Driving Automation

Automation Level	Short Name	Longitudinal/ Lateral Control	Driving Environment Monitoring	Fallback Situation Control	System Capability/ Driving Modes
0	No Automation	Human	Human	Human	None
1	Driver Assistance	Human & System	Human	Human	Some
2	Partial Automation	System	Human	Human	Some
3	Conditional Automation	System	System	Human	Some
4	High Automation	System	System	System	Some
5	Full Automation	System	System	System	All

Source: Adapted from "Definitions for terms related to driving automation systems for on-road motor vehicles", by SAE International, 2016, SAE Standard J3016. ©2016 SAE International.

happen and is responsible for constantly monitoring the vehicle and manually overriding the system wherever necessary. This is the level that the bulk of today's vehicles have reached.

Level 2, or *Partial Automation*, is the first level at which the vehicle is capable of moving on its own, with the system taking over both lateral and longitudinal control in defined use cases. Again, the driver must constantly monitor what is happening and be ready to manually override the system at all times. *Level 3*, or *Conditional Automation*, is a significant step up. As well as taking over both lateral and longitudinal control, the system is also capable of recognizing its limits and notifying the driver. The driver is no longer required to monitor the drive, but must be ready to manually override within a given timeframe if the system requests it.

Level 4 is *High Automation*. The system can take over all driving operations within a defined use case, and the driver has no part to play in either monitoring or back-up. Finally, *Level 5*, or *Full Automation*, sees the system take over the entire dynamic driving task in all use cases. This entails a fully functional vehicle that is capable of driving from a starting point to the destination without any intervention on the part of its occupants.

1.3   What benefits does SDV technology offer?

Reaching SAE Level 5 would make roads safer for everyone. Autonomous vehicles do not get tired or distracted, they don't drink alcohol, and they don't violate traffic regulations. Compared to humans, they also think faster. All these factors could reduce the number of accidents, deaths, and injuries on the road by eliminating human error. NHTSA figures suggest that 94% of accidents are caused by drivers [5], so self-driving vehicles clearly have the potential to improve safety. Fewer accidents should also mean lower insurance premiums.

Driverless cars would also free up time for vehicle occupants. Now all occupants are passengers, meaning they can catch up on chores, make use of entertainment systems, or take a nap while the vehicle drives them to their destination.

Eliminating the human factor will also lead to optimized driving behavior and improved traffic flow, easing congestion. This will offer both financial and environmental benefits, as well as reducing the risks to people's health, especially when combined with electric vehicle technology. According to the US-based Union of Concerned Scientists [6], general transportation is responsible for over half of carbon monoxide and nitrogen oxide air pollution. While many self-driving cars might still emit these same pollutants, their improved efficiency would mark a major step forward to a cleaner future.

Ultimately, the question of whether SDV technology will be better or worse for the environment will depend on technological and policy choices that have yet to be made. Automated vehicles could reduce energy consumption in transportation by as much as 90%, or increase it by more than 200%, according to research by the Department of Energy (DOE) [1]. That difference matters; more than a quarter of greenhouse gas emissions come from the transportation sector, according to the Environmental Protection Agency (EPA). "The impacts if you look into full vehicle automation could be huge," says Jeff Gonder, a transportation researcher at the National Renewable Energy Laboratory (NREL), "hugely positive or you could have huge increases in energy use" [2].

Improvements in technology in the field of autonomous vehicles will assist growth in other fields, too. Driving will take on a new meaning, changing people's perceptions of roads, commuting, and travel in general. Ride-hailing apps such as Uber and Lyft have already disrupted the taxi status quo, and now SDV technology promises to push the boundaries even further. Self-driving applications are not limited to cars but could also have far-reaching implications for sectors such as agriculture, delivery companies, and even security.

1.4 Why do we need another book about autonomous cars?

At the time of writing (2019), there are a number of books that describe the impact autonomous vehicles may have in the future (social, legal, etc.), but very few that provide any in-depth explanation of SDV technology. The primary goal of this book is to explain the technical aspects of building an autonomous vehicle. The book features numerous practical examples distilled from the author's professional experience in this area and a wide array of other resources.

Ultimately, this book is an attempt to bridge the gap between non-technical books that cover the broader implications of autonomous vehicles, and in-depth technical handbooks written for engineering professionals.

1.5   Whom is this book aimed at?

This book is intended for technology enthusiasts and anyone else who is interested in understanding how SDVs are developed. You could be an enthusiastic follower of the latest innovations, or someone who is keen to get involved in this fast-paced industry as a student, industry professional, innovator, start-up founder, or investor. The opportunities in this emerging field are not limited to the ambit of the vehicles themselves, but also encompass SDV enablers such as policymakers, insurance firms, infrastructure companies, cybersecurity specialists, mechanical engineers, electrical engineers, computer scientists, and many more. Last, but not least, this book will be of great assistance to technology journalists seeking a quick overview of the field.

Building an SAE Level 5 SDV is like completing a jigsaw puzzle with thousands of pieces. Many of the gaps have been filled by technology advancements in recent years, but parts of the puzzle are still missing. Some of the pieces are technical in nature, but others involve social, legal or ethical issues. Finishing the puzzle will require people from various disciplines. On the technical side alone, this will include mechanical engineers, electrical engineers, software developers, cybersecurity specialists, and computer scientists.

1.6   How is this book structured?

For simplicity, the core focus of this book is divided into three sections: hardware, software, and what we refer to as 'putting it all together'. Chapter 2 examines the different types of hardware available for use in an autonomous vehicle, discussing why each element is required and how it works. Chapters 3 and 4 look at the software components, and chapter 5 discusses how the two threads come together to create an autonomous vehicle. Chapter 6 takes a broader look at related issues such as back-end systems and cybersecurity, and Chapter 7 rounds things off with an overview of the strategies and applications that companies are pursuing in this field at the present time.

1.7 Disclaimer

This book is written by a practitioner in the field. That means it provides you with a level of technical knowledge that could be used as a basis for creating a self-driving vehicle.

It is important to remember that vehicles such as passenger cars are extremely complex technical products, the result of millions of person hours of professional development and testing. Working with or handling vehicle components could pose a risk of serious injury or death, and should only be done by qualified technicians.

Operating modified vehicles without a special license is illegal in some countries, and may potentially cause harm to you or others. In Germany, for example, all new types of vehicles or custom modifications to series production vehicles require explicit technical approval from official certification bodies such as the Technischer Überwachungsverein (TÜV).

Any custom modifications to the vehicle will almost certainly void your vehicle's warranty, and may cause permanent damage to the vehicle.

Furthermore, testing SDVs on public roads typically requires special permission from local transportation authorities. Many countries and states prohibit the testing of such vehicles on public roads.

Please remember that this book does not provide an exhaustive list of the technology required to develop safe SDV prototypes, let alone production-grade vehicles. Many details have been intentionally left out for the sake of brevity.

References

[1] Austin Brown, Brittany Repac, and Jeff Gonder. Autonomous vehicles have a wide range of possible energy impacts. Technical report, NREL, University of Maryland, 2013.

[2] Justin Worland. Self-driving cars could help save the environment - or ruin it. http://time.com/4476614/self-driving-cars-environment/. [Online; accessed 08-Jan-2018].

[3] Flynn PL. Leonardo da vinci's car. http://www.leonardodavincisinventions.com/mechanical-inventions/leonardo-da-vincis-car/. [Online; accessed 20-May-2018].

[4] Taxonomy SAE. Definitions for terms related to driving automation systems for on-road motor vehicles. *SAE Standard J*, 3016, 2016.

[5] Santokh Singh. Critical reasons for crashes investigated in the national motor vehicle crash causation survey. Technical report, National Highway Traffic Safety Administration, 2015.

[6] UCS. Vehicles, air pollution, and human health. http://www.ucsusa.org/clean-vehicles/vehicles-air-pollution-and-human-health. [Online; accessed 24-Dec-2018].

[7] Wikipedia contributors. General motors motorama — Wikipedia, the free encyclopedia. https://en.wikipedia.org/wiki/General_Motors_Motorama, 2019. [Online; accessed 07-Nov-2018].

Chapter 2

Hardware

In this chapter we will be looking at the hardware elements that make fully autonomous driving possible, divided for simplicity into three main categories: sensors, the computing platform, and the actuator interface.

The mechanical components that make up standard vehicles are not the focus of this book, so this chapter does not include descriptions of systems such as the engine, transmission, power train, and suspension. For the sake of brevity, we shall simply assume that each of these parts or systems is controlled by one or more electronic control units (ECUs) that are responsible for ensuring its safe and proper function. For example, when the actuator interface (described at the end of this chapter) sends a steering wheel command to turn the steering wheel several degrees to the left, it assumes that the ECU that is responsible for the steering wheel actuator will interpret the command correctly, carry out a series of internal actions to execute the command, and control the end output.

Our interest in this book is in examining the additional technologies that make self-driving cars a reality. Let us start with the eyes and ears of autonomous driving: sensors.

2.1   Sensors

The first step in creating a self-driving vehicle is to make it aware of its surroundings, and sensors are undoubtedly the most important way to achieve this. There are basically two types of sensors: passive and active.

Passive sensors work by absorbing ambient energy, while active sensors emit some form of energy into the environment and receive and measure the reflected signals. For example, cameras are passive, while radar and lidar are active. The ability of active sensors to operate in different environmental conditions depends on their specific mode of operation; for example, lidars work in the dark, but cameras typically do not, even though both rely on light energy. Regardless of which technology they use, active sensors must also be capable of dealing with noise and interference from the environment.

As well as perceiving the external environment, self-driving vehicles also need to measure their internal conditions. Sensors that measure a vehicle's internal state are called *proprioceptive sensors*, while those that allow the vehicle to 'see' outside are called *exteroceptive sensors*. An SDV typically uses both types of sensors to calculate the vehicle's position relative to its surroundings. Nevertheless, it is important that an SDV has the ability to calculate its position based solely on proprioceptive sensor readings, since information from exteroceptive sensors may not always be available.

Each sensor has its own unique properties that determine its level of complexity. Some sensors are easy to incorporate in vehicles because they are designed to serve a single purpose, and hence have limited capabilities and complexity. Tactile, motor, and heading sensors generally fall into this category. Other sensors are more sophisticated, and require the use of algorithms to exploit the full potential of the information they provide. These are typically sensors that rely on forms of energy that are more difficult to control and detect. This category includes active ranging sensors (ultrasonic sensors), motion/speed sensors (Doppler radars), and vision sensors (cameras). Harnessing their complexity comes with a pay-off: they provide a broader range of rich and varied information and have a wider range of operation. Some of them can even be used for multiple purposes, e.g., simultaneous ranging and detecting.

SDV sensors collect data and pass it on to a computing platform, which analyzes the data and determines what actions the vehicle should take next. Generally speaking, sensors are only part of the equation. They typically need to be accompanied by a set of software components and tools that process the raw sensor data to give it meaning, enabling the vehicle to exploit the information for its decision-making processes. This combination of supporting software components and tools forms what we call the *middleware*, the intermediate layer of the system architecture that bridges the gap between the hardware (sensors and actuators) and the SDV algorithms. Middleware and SDV software in general will be discussed in more detail in Chapter 4.

In this context, SDV developers face a three-pronged task: firstly, to design an optimum sensor configuration that takes into account functionality, cost, vehicle design, and other factors; secondly, to implement SDV algorithms and other tools that are capable of processing this data, and making the best possible decisions to achieve the defined objective; and, thirdly, to instruct the vehicle platform to perform the desired action. This, in simple terms, is the route map to creating a successful autonomous vehicle. It may seem like a daunting task, but the ability to achieve it ultimately stems from a clear understanding of how each individual component works.

2.1.1 Key considerations

With so many different types and models of sensors available, how do we go about choosing the best configuration for an SDV?

One approach is to observe the world around us. Nature teaches us that every organism has the set of organs it needs to survive and reproduce in its natural habitat. Bats are active at night, so they rely on sonar-like echolocation to sense their environment. In contrast, eagles use their exceptional eyesight to spot prey at enormous distances, yet an eagle's eyes are practically useless in the dark. Similar principles apply to the task of choosing sensors for an SDV, requiring us to carefully consider the conditions and environments in which the vehicle will operate. For example, long-range radars may be essential for self-driving private passenger cars speeding down a freeway, but, arguably, are much less suitable for last mile delivery SDVs that need to sense as many obstacles as possible in their immediate vicinity. Similarly, Global Navigation Satellite System (GNSS) sensors are useful for SDVs that operate outdoors, but serve little purpose for indoor SDVs.

Sensor configuration is not just a question of functionality, however, but also of cost. Some sensors, such as lidars, are still relatively expensive, which may rule them out for lower-end vehicle models. If the cost is prohibitive, then it may be necessary to compensate for the lack of one particular type of sensor by making intelligent use of other, more readily available, sensors. Whatever types of sensors we ultimately choose, it is essential to ensure that the vehicle has enough computational power to process data from all the sensors simultaneously.

Another key issue is vehicle design and aesthetics, especially in the self-driving passenger car segment where aesthetic considerations still play a major part in customers' purchasing decisions. The challenge of choosing the right sensors and placing them in the vehicle without compromising design quality is a classic trade-off between form and function.

2.1.2 Types of sensors

Ultimately there is no such thing as the perfect sensor, so developers normally use a combination of various sensor types, as shown in Figure 2.1. Even within specific sensor categories, there are subtle differences between different brands and models that may have to be taken into account. A good understanding of each sensor's strengths and limitations is therefore a crucial prerequisite for choosing the best option in each case. Let us take a closer look at the various sensors that are commonly used in autonomous vehicles.

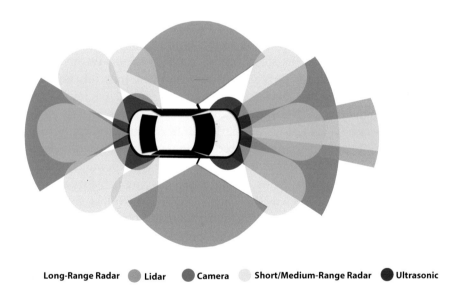

Long-Range Radar ● Lidar ● Camera ● Short/Medium-Range Radar ● Ultrasonic

Figure 2.1: Example of SDV sensors configuration.

2.1.2.1   Radars

Radar, which stands for radio detection and ranging, is a sensor technology that employs radio waves—i.e., electromagnetic wavelengths longer than infrared light—to detect and track objects. The first serious experiments using radio waves to detect distant objects were conducted in the early 1930s, but the new technique really gained momentum during World War II when both the Allied and Axis powers realized how much potential the new technology had for military applications.

In the late 1980s, Toyota pioneered the deployment of automotive radars in their vehicles [17]. Since then, other car manufacturers have embraced the technology and developed it further. The development of 77 GHz (and, more recently, 79 GHz) radar to complement earlier generation 24 GHz technology has led to better accuracy and resolution, both of which are essential for operating SDVs safely and reliably. Radar has become one of the most widely deployed sensors in modern vehicles, playing a key role in a number of advanced driver assistance systems (ADAS), including adaptive cruise control (ACC), blind spot monitoring, and lane change assistance. Figures 2.2(a) and 2.2(b) show examples of automotive radars.

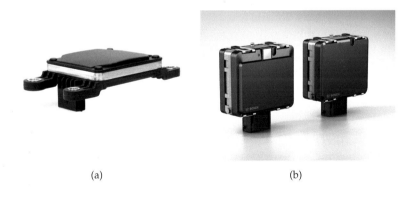

(a)   (b)

Figure 2.2: (a) Continental long-range radar (Reprinted with permission from Continental AG. ©2017 Continental AG). (b) Bosch mid-range radar. (Reprinted with permission from Bosch Media Service. ©2016 Robert Bosch GmbH)

The anatomy of radars

Radar technology makes use of echoes, emitting pulses of radio waves that are then reflected off surrounding objects. The returning waves provide information on the direction, distance, and estimated size of each object. Radars can also be used to determine the direction and speed of an object's movement by releasing multiple consecutive pulses.

There are two basic types of radar: echo and Doppler. Echo radars work as described above. By obtaining data from two or more echo radars placed in different positions, it is also possible to capture additional information on an object's position, such as its angle. A Doppler radar further enhances this capability by analyzing wave phases. It does this by keeping track of each particular wave and detecting differences in the position, shape, and form of the wave when it returns. This information can then be used to determine whether the wave has undergone a positive or negative shift. A negative shift means that the object is most likely moving away from the radar, while a positive shift indicates that it is moving towards the radar. The amount of shift can be used to determine the object's speed.

Pros and cons of radars

Thanks to its long range coverage and Doppler functionality, radar has become the primary sensor for detecting and tracking distant objects. It offers a number of key advantages. For example, radar can be utilized in any light conditions (including direct sunlight and darkness), in any disruptive weather conditions (for example rain, fog, and snow), in windy locations, and when traveling at high speed. Radars also offer adequate resolution even at longer ranges (up to 250 meters), and are available at a reasonable price in volume production (though some other types of sensors are cheaper). Last but not least, both the position and the velocity of detected objects can be estimated due to the Doppler effect.

The downsides of radar include poorer results with non-metallic objects, and its relatively narrow opening angle. Some radars come with features that enable the opening angle and range to be adjusted dynamically depending on the vehicle's velocity, as shown in Figures 2.3(a) and 2.3(b). When the vehicle is driving at high speed, the system reduces the opening angle to get the maximum range. When it is traveling at a lower velocity (e.g., in city traffic), the opening angle is set to its maximum by reducing the range, allowing better detection of pedestrians, bikes, and other objects close to the vehicle.

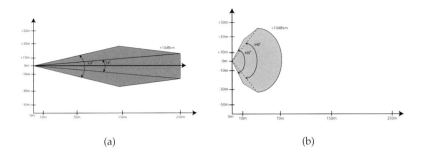

(a)   (b)

Figure 2.3: Radar operating angle for long range (a) and short range (b). The wider the opening angle, the shorter the operating range. (Adapted from "ARS 408-21 Premium Long Range Radar Sensor 77 GHz Datasheet" by Continental. ©2017 Continental AG)

2.1.2.2   Lidars

Lidar is an acronym for light detection and ranging. Lidar technology is based on the same principle as radar; in other words, it determines the location and distance of objects based on the reflection of transmitted energy. However, lidar works using pulsed laser light instead of radio waves.

Since they were invented in the late 1950s, lasers have been harnessed for a diverse and steadily increasing range of applications. The use of lidar in self-driving vehicles is just one example of this trend. Unique characteristics such as high resolution and non-metallic object detection have made lidar a popular choice for 3D surveying and mapping applications. This dovetails perfectly with SDV development, which relies on the availability of high definition maps for precise localization and navigation (as we will see in Chapter 3). Figure 2.4 shows some examples of the lidar sensors.

Figure 2.4: Velodyne Alpha Puck, Velarray, and VelaDome lidar sensors. (©APJarvis, https://commons.wikimedia.org/wiki/File:Velodyne_AlphaPuck_ W_Velarray_VelaDome_Family_BlueLens.png, https://creativecommons.org/ licenses/by-sa/4.0/legalcode)

To ensure that the laser pulses emitted by lidar devices do not damage people's eyes, the energy of automotive lidar beams is strictly limited to the eye-safe level of Class 1 laser products [15]. Other applications use much higher power beams. In aerial surveying, for example, the pulses need to have enough energy to penetrate tree canopies and reach the forest floor.

The anatomy of lidars

As mentioned above, lidars work in a similar way to radars. Whenever a transmitted laser pulse hits an object, the pulse is reflected back to the sensor. The object distance can then be calculated by measuring the pulse travel time.

The difference is that lidars use light waves. Lidar sensors are capable of firing rapid pulses of laser light at rates of up to several hundred thousand pulses per second. Modern lidars are also capable of transmitting multiple vertically aligned pulses, or 'channels', in a single scan, providing information on an object's height. This may be useful for certain perception algorithms such as noise filtering and object recognition.

Lidar sensors essentially consist of three main components: laser diodes to generate the laser beams, photodiodes to receive the returning (reflected) signals, and a servo-mounted mirror to direct the laser beam horizontally and vertically. The returning signals are captured by the photodiodes and processed by the sensor's signal processing unit. The sensor outputs the detected objects as a series of point cloud data, with each pixel representing the measured distance and exact location in 3D coordinates relative to the sensor as depicted in Figure 2.5. More intelligent lidars also return a list of recognized objects such as car, pedestrian, and so on.

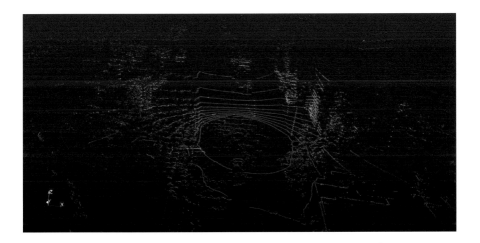

Figure 2.5: Example of raw sensor data (3D point cloud) visualization.

Figure 2.6 shows the key components of a lidar and illustrates how they work. The generated laser pulses are guided through a mirror rotated by a servo motor. This mirror can be tilted to transmit pulses at different vertical angles. An optical encoder provides feedback to the servo motor to enable precise control of the mirror and the resulting laser transmission. The returning signals are captured by a detector (typically an array of photodiodes) and are processed by the sensor's signal processing unit.

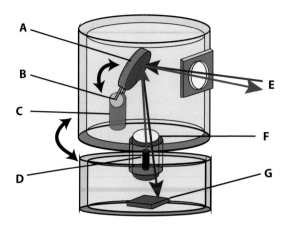

A : Tilting Mirror   B : Optical Rotary Encoder   C : Servo Motor   D : Laser Source
E : Objects   F : Optical Rotary Encoder   G : Receiver

Figure 2.6: (Simplified) principle of lidar operation. (Adapted from "Optical encoders and LiDAR scanning" by Renishaw. ©2019 Renishaw plc)

Pros and cons of lidars

With their small beamwidth and relatively long range coverage, lidars have become the sensor of choice for high resolution 3D mapping. Lidars also play an important role in indoor positioning, and in other areas where satellite-based GNSS is unavailable. By measuring the intensity of received infrared light, lidars also have the potential to be used as a reliable day/night detector, as the sun transmits significantly more infrared light than the generated laser beams in the daytime [5].

Despite lidar's usefulness in automotive applications, the high unit cost of lidar sensors remains the biggest hindrance to broader adoption of this technology in serial production vehicles. However, the ongoing development of solid-state lidars (i.e., lidars that are non-rotating and have no moving parts) promises to achieve a significant reduction in sensor cost and size.

Because laser beams also reflect off small particles such as fog, and dust, lidars are more sensitive to their environment, and generate more 'noise' in bad weather than their radar counterparts. This makes it more complex to incorporate lidar sensors in vehicles than radars. Although filtering algorithms can sometimes help reduce the interference caused by snowflakes or raindrops, they are much less effective if the laser pulses are obstructed by dust, ice or snow on the surface of the sensor itself. This problem can be avoided by placing the sensor behind the windshield, but locating it inside the vehicle and within the visual range of the windscreen wipers makes 360-degree perception impossible, and may cause conflicts with other sensors, such as vision cameras and rain sensors [12].

2.1.2.3   Ultrasonic sensors

Ultrasonic refers to acoustic waves beyond the human audible range, i.e., frequencies above 20 kHz. As their name implies, ultrasonic sensors use these high frequency acoustic waves for object detection and ranging. Animals, such as bats, use similar principles to detect and locate their prey in low-light environments.

For decades, ultrasonic sensors have been used for monitoring and diagnostics in medicine, the maritime sector, and other industries. But it wasn't until the 1980s that they entered the automotive industry with the introduction of Toyota's ultrasonic-based parking assistance systems, which were swiftly embraced by other car manufacturers.

Nowadays, ultrasonic sensors in vehicles offer more than just parking assist capabilities. For example, they have also been harnessed for gesture recognition as part of human machine interface (HMI) systems that enable touch-free control of head unit and infotainment systems, allowing users to navigate through media tracks and adjust playback. Figures 2.7(a) and 2.7(b) show an example of automotive ultrasonic sensors and its application as side parking sensor, respectively.

(a)   (b)

Figure 2.7: (a) Bosch ultrasonic sensors. (Reprinted with permission from Bosch Media Service. ©2016 Robert Bosch GmbH). (b) Example of ultrasonic sensor as side parking sensor. (©Basotxerri, https://commons.wikimedia.org/wiki/File:VW_Golf_VII_-_Parking_sensor_02.jpg, https://creativecommons.org/licenses/by-sa/4.0/legalcode)

The anatomy of ultrasonic sensors

Like lidars and radars, ultrasonic sensors are classed as time-of-flight sensors, which means they work by transmitting packets of waves and calculating how long it takes for the waves to return. It is essential that the acoustic waves used in ultrasonic sensors are non-audible to humans, because the waves must be transmitted with high amplitude (>100dB) for the sensors to receive clear reflected waves.

The sensors essentially comprise a transmitter, which converts an electric alternating current (AC) voltage into ultrasound, and a receiver, which generates AC voltage when a force is applied to it. Depending on the material used, both functions can be combined into a single transceiver.

Due to the wide opening angle of the sound waves—and assuming optimum placement of the sensors in the vehicle—it is possible to determine the exact position of detected objects using the trilateration of overlapping signals, using the same principle as satellite-based positioning, as shown in Figure 2.8.

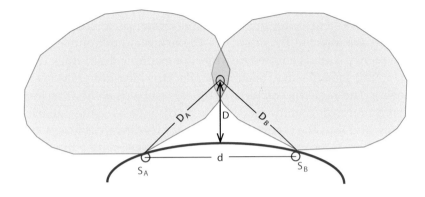

Figure 2.8: Ultrasonic trilateration. The distance to the detected object (D) can be deduced by applying Pythagoras' law to the distance of the object to each sensor (D_A and D_B) and the distance between the two sensors (d). (Adapted from "Ultraschallsensorik", by Martin Noll, Peter Rapps, Hermann Winner (ed.), Stephan Hakuli (ed.), Christina Singer (ed.), 2017, Handbuch Fahrerassistenzsystem, ATZ/MTZ-Fachbuch, p. 253. ©2015 Springer Fachmedien Wiesbaden)

Pros and cons of ultrasonic sensors

Due to their relatively affordable price, ultrasonic sensors are typically deployed as an economical means of detecting the presence and position of objects in the vicinity of the vehicle, for example in parking assist applications. Ultrasound performs well in most situations, since it is not affected by the majority of weather conditions. Although ultrasonic sensors provide relatively little detail, they are not dependent on light, which can be an advantage in cases where not enough or too much light could yield misleading results. They also work in rain, fog, and snow, as long as the sensor itself is not covered by dirt, snow, or ice. In indoor, city, or crowded environments, their ability to detect non-metallic materials offers an additional safety measure in regard to pedestrians.

There are a few concerns that need to be addressed when using ultrasonic sensors, however. They typically have a low resolution and low operating range, and they exhibit limited functionality in situations involving high wind speed or high velocity, which tend to be exactly the moments where optimum sensor performance is crucial. What's more, they are affected by sounds in the external environment. Other high-frequency sounds emitted nearby—for example, train-track friction sound, or 'rail squeal'—may negatively affect the measurements taken by the sensors. The angle and material of reflective objects also has an impact on the echo received. As the angle of the wave increases, the intensity of the waves, and thus the accuracy of the sensor readings, decreases. Incorrect calculation of the waves could cause errors in estimating the distance or presence of other objects around the vehicle.

2.1.2.4 *Cameras*

Camera technology has been around for a long time and is undoubtedly one of the most remarkable innovations in human history. The invention of digital cameras in the late 1980s disrupted the traditional business of analog photography. Today's digital cameras are inexpensive, and are available almost everywhere, thanks to their inclusion in smartphones.

The use of camera technology in the automotive sector was first explored by General Motors in its 1956 Buick Centurion concept car. The Centurion featured a rear-mounted television camera (also referred to as a 'backup camera'), which sent images to an in-vehicle TV screen. It wasn't until three decades later, however, that Toyota paved the way for the first serial production of vehicles with integrated rear-view cameras [6]. Nowadays, camera technology is used not only to provide a live rear-view image stream to assist parking, but also as a key enabler for various ADAS innovations, from lane departure warning and speed limit information to augmented reality and autopilot. In some countries, cameras are taking on an even more significant role in the auto sector. In the U.S., for example, rear-view cameras are mandatory in all new vehicles built from May 2018 onwards.

Figures 2.9(a) and 2.9(b) show an example of automotive stereo cameras and a mounting example of front-looking cameras inside the vehicle, respectively.

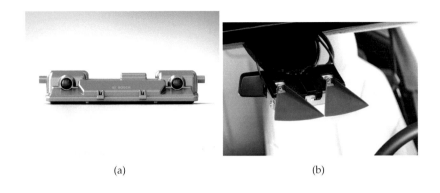

(a)   (b)

Figure 2.9: (a) Bosch stereo video camera. (Reprinted with permission from Bosch Media Service, ©2017 Robert Bosch GmbH). (b) Stereo video camera mounted on the windshield. (Reprinted with permission from Bosch Media Service. ©2017 Robert Bosch GmbH)

The anatomy of cameras

Cameras, unlike lidars, radars, or ultrasonic sensors, are passive sensors. They passively receive light waves and do not actively transmit any form of energy (time-of-flight cameras are an exception which we will discuss below). Cameras essentially consist of three main components: optics, an image sensor, and the image processor.

The design of the objective lenses and filters that make up the cameras' optics depends on the intended application. Front-view cameras typically employ long-focus lenses with a large aperture to enable the camera to 'see' as much as possible in low-light situations, whereas side-view and rear-view cameras use wide-angle lenses with small apertures to capture as many nearby objects in the surrounding environment as possible. The image sensor and the image processor are responsible for capturing, filtering and processing the received light waves into digital raw video streams that can be transmitted via low-voltage differential signaling (LVDS) or Ethernet interfaces. Some intelligent camera systems also feature a powerful digital signal processor (DSP) that performs real-time object detection, sign recognition, lane detection, and other tasks, and transmits the detected object lists as separate bus messages.

Stereo cameras are basically two mono cameras facing in the same direction. This arrangement has two separate input video streams, one for the left camera and one for the right. Figure 2.10 shows an example of depth calculation in stereo camera.

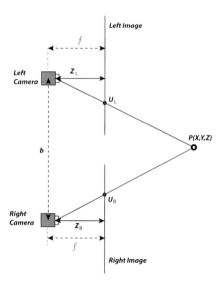

Figure 2.10: Depth calculation in stereo cameras. Assuming both left and right cameras have the same focal length f, the real-world point P is projected as u_l and u_r, respectively. The distance between two projected points is referred to as *disparity* and is calculated as $f * b/z$, where b is the distance between the two cameras. Thus z, or the distance from the cameras to the real point P, can be determined using the formula: $z = f * b/disparity$. (Adapted from "Optical encoders and LiDAR scanning" by Renishaw. ©2019 Renishaw plc)

One of the key properties of stereo cameras is the ability to carry out correspondence search, which involves searching for similarities in different sensors to create a continuous image of the surroundings. Various algorithms can be used to do this. Area-based algorithms consider a small area of one image and look for a similar area in another image. In contrast, feature-based algorithms identify unique features in each image in order to match up common points. Rather than computing an area, the calculations can be drawn from much smaller identifiable components of the image, including edges, corners, and lines. *Epipolar geometry* can be used as a basis for reducing the complexity of correspondence searches, as long as there are several pixels in one image that correspond to pixels in the other image. Correspondence can be generated using *epipolar lines*, a process in which an object in one camera view is projected onto the image plane of the second camera. This feature-based technique allows the system to identify similarities, and combine the images. A similar principle is used to generate a surround view (360 degrees) from four or more fisheye cameras with overlapped horizontal fields of view (FOV).

Time-of-flight (TOF) cameras are able to capture an entire 3D scene from a fixed position. To determine the distance of signals, TOF cameras usually have a photonic mixer device (PMD) attached next to the lens. An infrared lighting source is employed to determine the distance of objects based on the time-of-flight operating principle used in lidars, radars and ultrasonic sensors. This separate distance measurement makes distance calculations in TOF cameras far simpler than in stereo cameras.

Pros and cons of cameras

Compared to range sensors such as lidars, radars and ultrasonic sensors, cameras capture a wider gamut of frequencies, including colors. This allows for a richer semantic interpretation of scenes, including lane detection, traffic sign recognition, and so on. In some cases, visual-based localization using cameras may yield better results than lidar-based localization, for example, in situations where landmarks such as buildings are easier to distinguish by their texture rather than just their structure. A further advantage is that cameras typically have a lower unit cost than radars and lidars.

Camera technology is, however, sensitive to ambient light and weather conditions. Cameras do not work well in direct sunlight, and their effectiveness is also significantly impaired in poorly lit environments. Changes in the weather also tend to have a negative impact, with heavy rain, snow, or foggy conditions all potentially rendering camera images useless.

2.1.2.5   Global Navigation Satellite Systems

Much like radar, a Global Navigation Satellite System (GNSS) has its roots in military applications. The first fully functional GNSS was Navstar, the predecessor of the United States' modern-day global positioning system (GPS). Navstar, which stands for navigation system using timing and ranging, was invented in the 1970s by the U.S. Department of Defense to provide a quick and effective means of locating U.S. military units anywhere in the world, particularly the newly created missile-launching submarines operated by the U.S. Navy [9]. The major importance of such global navigation systems for both military and civil applications has led to several other satellite constellations being launched. These include the European Union's Galileo, Russia's GLONASS, and China's BeiDou systems.

Readily available satellite-based navigation with global coverage gave rise to several key innovations that have paved the way for SDV development. Aided by a digital map and an on-board computer, vehicles can automatically locate themselves on a global map, calculate routes, and navigate drivers to their chosen destination. In the early 1980s, the Japanese company Honda became the first car manufacturer to launch a commercial car navigation system with a map display. Nowadays, drivers benefit from real-time traffic information services such as Google Maps, which provide drivers with instant updates on alternative routes to help them avoid tailbacks many kilometers ahead.

In terms of its specific use in SDV navigation, it is important to remember that GNSS has limited accuracy and cannot be relied on in every situation. Nevertheless, GNSS will undoubtedly continue to play an important role in SDV development for many years to come.

The anatomy of GNSS

GNSS relies on satellites spread throughout the sky in different orbital planes. In the case of GPS, global coverage requires a constellation of at least 24 operational satellites continuously transmitting signals back to Earth, including the satellite's ID, current time, and location.

GNSS receivers are passive and exteroceptive. In the case of GPS, a receiver needs signals from three or more different satellites to fix its location on the ground. Only three satellites are required if the GPS receiver has a built-in atomic clock. However, most GPS receivers have simpler clocks, so signals from at least four satellites are normally needed to compensate for timing inaccuracies.

The position of a GNSS receiver is calculated based on a mathematical principle called *trilateration*. The method works by first calculating the propagation time of each signal, in other words the time difference between the time the signal left the satellite and the time it was received. Distance to the satellite can then be calculated by multiplying the propagation time by the speed of signal travel, which is equivalent to the speed of light. The position of the receiver is pinpointed as the area where all the signals intersect as depicted in Figure 2.11. By knowing the location of satellites A, B, and C as well as the distances to each of them, i.e., D_A, D_B, and D_C, the global position of the receiver can be determined.

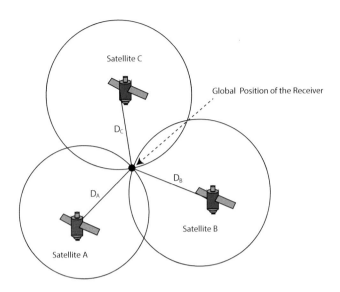

Figure 2.11: Principle of trilateration for determining the global position of a GNSS receiver.

Pros and cons of GNSS

A GNSS system with global coverage, such as GPS or GLONASS, can be used to determine the position of a suitable receiver anywhere on the surface of the Earth. GNSS is based on absolute positioning, which means it does not suffer from the accumulated error that affects *inertial measurement units (IMUs)* and *odometry sensors* over long periods of use. GPS receivers have become commonplace and affordable, and are now available almost anywhere thanks to their ubiquitous use in modern smartphones.

One of the biggest drawbacks of GNSS-based positioning is that it requires a clear line of sight between the receiver and the satellites to work reliably. That means GNSS-based positioning works best in open areas with unobstructed views and does not work at all in indoor areas such as garages and tunnels. In certain environments, for example dense urban areas with tall buildings packed closely together, GNSS signals suffer from multipath propagation. This occurs when a signal is reflected off other objects in the environment and reaches the receiver via several different paths, potentially leading to a significant degradation of positioning performance. Another issue is that publicly accessible GPS can only achieve a positioning accuracy of approximately 3 meters (1 meter for the ESA's Galileo system), which may not be precise enough for SDV applications. Location accuracy can, however, be significantly improved by using technologies such as the differential global positioning system (DGPS) or real-time kinematic GPS (RTK GPS), though both technologies require dedicated base stations in fixed locations and are therefore only available in certain areas of the world.

2.1.2.6 Inertial Measurement Units

As discussed in the previous section, GNSS requires a line of sight between the receiver and at least three or four satellites to work properly. SDVs may therefore need other techniques in places where satellite signals are not available, dense urban areas or indoor environments. *Inertial Measurement Units (IMUs)* are often used for this purpose.

IMUs enable a self-driving vehicle to determine its position and pose (the direction it is facing) by combining data from accelerometers, gyroscopes, and sometimes magnetometers. An IMU typically consists of three gyroscopes and three accelerometers that provide six *degree-of-freedom (DoF)* pose estimation capabilities (X, Y, and Z coordinates plus *roll*, *pitch*, and *yaw*). Some models also incorporate three magnetometers to deliver a nine-DoF pose estimation. Figure 2.12 shows some examples of IMU sensors.

Figure 2.12: Xsens MTi 1-series, 10 series, 100 series IMU sensor modules. (Reprinted with permission from Xsens. ©2019 Xsens Technologies B.V.)

The anatomy of IMUs

IMUs make use of inertial motion sensors, particularly gyroscopes. They identify the orientation of the vehicle based on a fixed reference frame. There are several types of gyroscopes. *Mechanical gyroscopes* use a spinning wheel or a fast-spinning rotor mounted on two gimbals and a support frame. Due to the angular momentum of the wheel, the original orientation of the wheel is constantly preserved regardless of changes in orientation of the support frame. Thus, the change of orientation can be deduced by measuring the angular displacements between the two gimbals with respect to an inertial reference frame. *Optical gyroscopes* are typically laser-based, and utilize a physical phenomenon called the *Sagnac effect* [14]. When two optical beams propagate in opposite directions in a rotating ring path, their propagation time, i.e., the time for the beam to return to its starting point, will differ fractionally; the beam that travels in the same direction as the rotation will take longer to return to the starting point than its counterpart, as shown in Figure 2.13. The angular velocity applied to the ring can be deduced by measuring the phase difference of the beams.

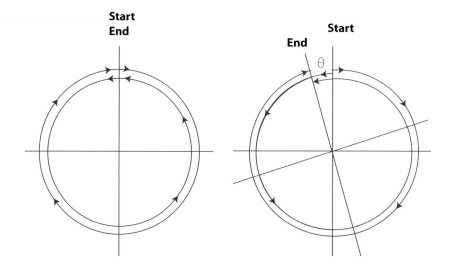

Figure 2.13: Sagnac effect. When two optical beams propagate in opposite directions in a rotating ring path (right), their propagation time will differ from the one when the ring is not rotating (left). By measuring the phase difference of the beams (θ) the angular velocity applied to the ring can be deduced. (Adapted from "Reflections on Relativity" by Kevin Brown. 2004. ©2004 Kevin Brown)

Another popular type is *Micro Electro-Mechanical System (MEMS) gyroscopes*. MEMS gyroscopes work on the principle of measuring the *Coriolis force*, an inertial force that deflects moving objects in a rotating system proportional to the system's angular velocity [13]. As illustrated in Figure 2.14(a), when a person is moving southward toward the outer edge of a rotating platform, that person has to increase the westward acceleration in order to maintain the southbound course. The increased westward acceleration is required to compensate the increasing eastward Coriolis force. A MEMS gyroscope containing a resonating mass that only moves along one direction (northward or southward) inside a frame with capacitive sensing elements on its both west and east sides. As the resonating mass moves northward, the platform's angular speed is determined by measuring the displacement of the resonating mass caused

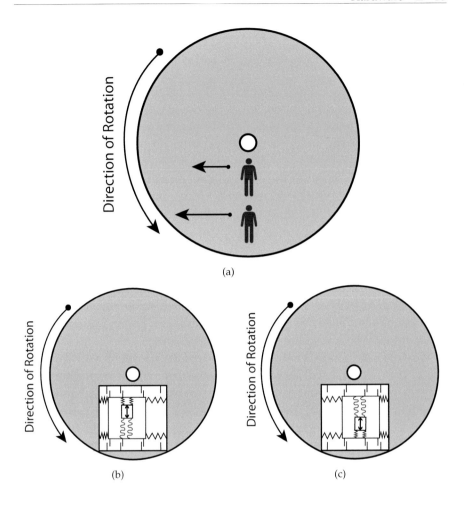

(a)

(b)　　　　　　(c)

Figure 2.14: Coriolis effect and the working principle of MEMS gyroscope. (Adapted from "MEMS Gyroscope Provides Precision Inertial Sensing in Harsh, High Temperature Environment", by Jeff Watson. ©2019 Analog Devices, Inc.)

by the westward Coriolis force, as depicted in Figure 2.14(b). Likewise, the angular speed is proportional to the displacement of the resonating mass caused by the eastward Coriolis force as the resonating mass moves toward the southbound direction as shown in Figure 2.14(c).

Due to their low power consumption, affordable cost, and high performance despite their small form factor, MEMS gyroscopes are widely used in modern electronic devices, ranging from consumer electronic appliances to safety-critical automotive and aerospace applications.

Another important component of IMUs are accelerometers, which work on the spring-mass-damper principle. The inertial force, damping force, and spring force can all be calculated to measure the force applied to the object. The pressure applied by the spring and mass are damped by the residual gas sealed in the device. There are various types of accelerometers including *capacitive accelerometers*, which measure the capacitance between a fixed structure and the proof mass, and *piezoelectric accelerometers*, which use a single crystal or piece of ceramic piezoelectric material and measure the voltage generated by this material when pressure is applied. Each accelerometer measures acceleration on a single axis, which is why IMUs tend to have three accelerometers to allow 3D calculations to be made.

Pros and cons of IMUs

IMUs are passive sensors designed to sense parameters that can be measured under any circumstances, such as the Earth's gravitational field and magnetic fields. Self-driving vehicles can always rely on getting information from IMUs because they are always available.

However, this high level of availability does not mean that IMUs are free of errors. Typical IMU errors include noise, offset and scale factor errors, which may be exacerbated by weather or temperature. Some IMUs include a barometer so that barometric pressure measurements can be used to account for pressure variations due to weather effects. In extreme situations with rapidly changing weather, the sea level barometric pressure may vary to an extent that temporarily compromises accuracy in the vertical direction.

2.1.2.7   Odometry sensors

Odometry sensors, or odometers, are special sensors designed to measure the distance traveled by a vehicle by multiplying the number of wheel rotations by the tire circumference. Odometers are either active (they need external power) or passive.

In the automotive industry, the odometer function is typically implemented in the form of wheel speed sensors. These provide information on current wheel velocity as well as the distance traveled by each wheel. Standard safety features, such as the anti-lock braking system (ABS), rely heavily on accurate wheel velocity information in order to work properly.

The anatomy of odometry sensors

Passive wheel speed sensors do not require an external power supply. A pulse wheel rotates synchronously with the wheel being monitored. The alternating teeth and gaps in the pulse wheel cause magnetic flux changes between the coil and the permanent magnet mounted in the head of the sensor. As shown in Figure 2.15, the sensor determines the speed by measuring the alternating voltage induced by the changes in magnetic flux.

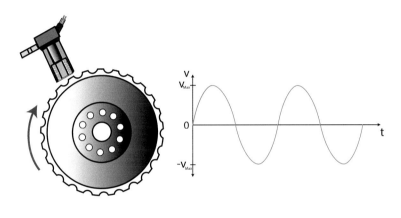

Figure 2.15: Working principle of a passive wheel speed sensor. The alternating teeth and gaps due to the wheel rotation (left) causes magnetic flux changes between the sensor's coil and permanent magnet and generates an alternating current (right).

Active wheel speed sensors work on a similar principle, but use a ring with alternating magnetic poles as the pulse wheel, as shown in Figure 2.16. The alternating pole change is detected by magnetoresistance or Hall sensors, and converted by the sensor into pulse-width modulated (PWM) signals. Unlike passive sensors, active sensors require an external power supply from their control unit.

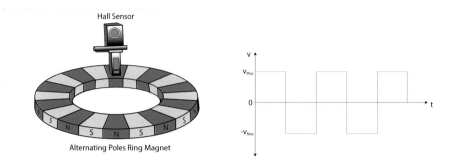

Figure 2.16: Working principle of an active wheel speed sensor. A Hall-effect sensor detects the alternating magnetic pole change due to the pulse wheel rotation (left) and the generated PWM signal is proportional to the rotation speed (right).

Pros and cons of odometry sensors

Odometers are inexpensive sensors that can provide accurate information on the distance traveled by a vehicle. Modern active sensors can detect wheel speeds as low as 0.1 km/h, and are largely unaffected by vibrations and temperature fluctuations [7].

However, they are susceptible to accumulated errors over time due to vehicle drift, wheel slippage, uneven surfaces, and other factors. This diminishes their accuracy over long distances [1]. Readings from odometry sensors are therefore normally combined with measurements from other sources such as GNSS and IMU to calculate a vehicle's position, using algorithms, such as the Kalman filter, to obtain an accurate overall result. Combining sensor data is discussed later in Section 3.5.

2.2 Computing platform

The sensors described in the previous section are the 'eyes and ears' of a self-driving vehicle, the technology that allows it to sense its environment. The computing platform is the 'brain' that makes sense of all this data, fusing the information from the sensors in real time to generate a view of its surroundings. Once the computing platform understands what is happening in the vehicle's environment, it can make decisions and send instructions to the *actuator interface* to perform whatever movements are required. This *sense-decide-act* loop lies at the heart of automated driving.

2.2.1 Key considerations

Building up a robust, highly accurate 3D image of a vehicle's surroundings is essential for safe and reliable autonomous driving, but it requires a level of computing power that can only be provided by a high-performance computer (HPC). Let us look at some of the key issues involved in choosing the best computing platform for an SDV.

- Data rate
  SDV sensors produce vast quantities of data, which need to be processed simultaneously. Depending on the number of cameras, lidars, radars, and other sensors in the vehicle, this may require data transfer rates of up to 1 Gigabyte per second. By way of comparison, that is roughly 1,000 times the data rate of HD video streams on YouTube. Not only does all this data have to be captured and processed, but it also poses a significant challenge in regard to storage. Access to stored raw sensor data is an essential part of fault diagnosis and other functions, so both the storage controller and the storage devices themselves (HDDs or SSDs) need to be designed to deal with these high data rates in order to avoid creating bottlenecks.

- Computing power
  As well as simultaneously processing huge quantities of data, the computing platform must also have enough computing power to make the right decisions in each and every situation. A few milliseconds of delay could have serious consequences, so near-zero latency is imperative. Take emergency braking, for example. At 100 km/h, a vehicle travels 28 meters every second. So if an SDV reacts one second too late while traveling at 100 km/h, the total braking distance is 28 meters longer than it should have been. At

140 km/h, that one-second delay would translate into an extra 39 meters of braking distance. In both cases, this could mean the difference between life and death. Just like drones and surgical robots, self-driving vehicles depend on the reliable delivery of actionable information in near-real-time.

■ Energy consumption
Experts predict that most SDVs in the future will be electric vehicles [16]. The maximum distance an electric vehicle can travel depends on the total energy consumption of all its electronic components as well as the drive train. To avoid an unnecessary drain on the battery, developers need to create computing platforms that offer maximum energy efficiency as well as high computing power.

■ Robustness
To ensure an SDV can operate safely in all possible geographical locations and climates—even in extreme temperatures—the computing platform needs to meet automotive-grade standards; for example, an operating temperature range from -40°C to 125°C. The computing platform and its hardware components must also be robust enough to withstand mechanical vibrations.

2.2.2 *Examples of computing platforms*

Due to high general interest (even from outside the traditional automotive industry) and growing research and development activities in the field of SDV technology, a number of technology companies have started to offer specially designed high performance computing platforms for SDV or other demanding applications.

Nvidia is well known for its graphics processing units (GPUs). Early on, Nvidia saw the potential of its GPUs to tackle perception challenges in autonomous driving, especially those that can be effectively addressed using deep learning, a topic we will discuss at greater length in Chapter 7. The NVIDIA DRIVE AGX Pegasus, as shown in Figure 2.17, is a scalable, powerful, and energy-efficient computing platform that has been specifically designed to facilitate autonomous driving.

Figure 2.17: NVIDIA DRIVE AGX Pegasus computing platform. (Reprinted with permission from NVIDIA. ©2019 NVIDIA Corporation)

Another company that has developed computing platforms for autonomous driving is TTTech Auto. The company specializes in robust networked safety controls in technology sectors ranging from computing to the aerospace industry. Developed according to ASIL D functional safety requirements, TTTech Auto's Renesas chipsets-based "RazorMotion" and Intel/Infineon chipsets-based "AthosMotion" (see Figure 2.18) prototype ECUs enable the development of SDV functionalities to series-production standards. Functional safety is discussed in more detail in Chapter 6.

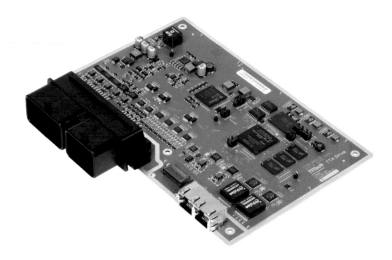

Figure 2.18: TTTech Auto AthosMotion computing platform. (Reprinted with permission from TTTech Auto. ©2019 TTTech Auto AG)

2.3 Actuator Interface

The actuator interface is responsible for translating the vehicle commands issued by the computing platform into the actual physical movement of the vehicle. For example, if the computing platform decides to turn the steering angle three degrees to the left, the actuator interface ensures that all the low-level commands for the steering control module (which may itself involve a closed-loop control system) are performed correctly and in a timely manner, such that the steering angle is turned no more and no less than three degrees to the left when the command is executed. The actual series of low-level controls needed to execute this command may vary from one vehicle to another. Thus, the actuator interface acts as a vehicle-independent abstraction layer that hides the complexity of each vehicle's specific low-level controls.

2.3.1 Components of an actuator interface

For the SDV to maneuver safely and reliably, the actuator interface needs to support both lateral and longitudinal controls, as defined by SAE International in SAE Standard *J670e Vehicle Dynamics Terminology* [10]. Lateral control involves the control of a vehicle's movements along the Y-axis or 'sideways'. Steering angle control falls in this category. Longitudinal control involves the control of a vehicle's movements along the X-axis. Acceleration pedal control and braking control are two examples that fall in this category. Figure 2.19 shows the vehicle axis system according to SAE J670e standard.

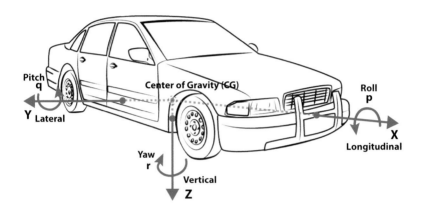

Figure 2.19: SAE J670 vehicle axis system. (Adapted from "Vehicle Dynamics Terminology: SAE J670e", by Society of Automotive Engineers Vehicle Dynamics Committee, 1978. ©1978 SAE International)

Because SDVs are controlled by software running on the computing platform, the vehicle's actuators must be fully programmable, or 'drive-by-wire-ready'. In a fully *drive-by-wire* vehicle, the driving functions are controlled by one or more electronic control units (ECUs) based on commands retrieved from the bus system as shown in Figure 2.20. To allow fully programmable lateral and longitudinal control, an SDV needs to have at least the following drive-by-wire components: steer-by-wire, brake-by-wire, and throttle-by-wire. *Steer-by-wire* allows lateral vehicle control using electronic steering commands (or messages) sent over the communication bus. *Brake-by-wire* and *throttle-by-wire* enable longitudinal vehicle control programmatically without any mechanical pedals. On receipt of braking or throttle commands, the responsible ECU translates these into an actual (physical) braking and acceleration action, respectively.

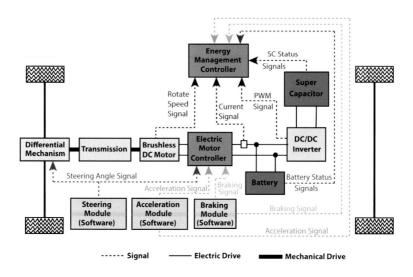

Figure 2.20: Example of a drive-by-wire architecture for electric vehicles. (Adapted from "Optimal Control for Hybrid Energy Storage Electric Vehicle to Achieve Energy Saving Using Dynamic Programming Approach", by Chaofeng Pan, Yanyan Liang, Long Chen, and Liao Chen, 2019, Energies 2019, 12(4):588. ©Chaofeng Pan, Yanyan Liang, Long Chen, and Liao Che, https://www.mdpi.com/energies/energies-12-00588/article_deploy/html/images/energies-12-00588-g001.png, "Architecture of the hybrid system", https://creativecommons.org/licenses/by/4.0/legalcode)

Because all vehicle control is performed by software, the control hardware for human drivers, such as steering wheel, acceleration pedal, etc., is no longer necessary. However, such hardware might still be there to allow a human driver to override SDV actions, if necessary. In level 5 SDVs, this hardware becomes completely obsolete as there won't be any driver in the vehicle that can override the vehicle's actions.

Drive-by-wire systems have been a key enabler for innovative ADAS applications such as adaptive cruise control and lane assist systems. At the same time, however, they also pose a significant risk of unauthorized vehicle control by paving the way for tampering and the injection of fake messages for the ECUs. Modern vehicles generally come with an extra layer of security to make it more difficult for hackers to manipulate the drive-by-wire system. Older vehicles are more vulnerable, however, and someone who knows how to interpret and manipulate the internal bus messages for the drive-by-wire system may be able to gain control of the vehicle.

Nevertheless, it is likely to be considerably easier to build an SDV prototype on the basis of an older vehicle than a modern series production vehicle. A further point to bear in mind is that the internal messages used for drive-by-wire systems vary from one car manufacturer to another, sometimes even between models from the same manufacturer, and are generally protected as proprietary data.

2.3.2   Enabling drive-by-wire systems

As mentioned in the previous section, an SDV requires fully programmable vehicle control by means of drive-by-wire or similar systems. The internal messages used in drive-by-wire systems are not generally publicly available and may be protected against manipulation.

However, several options may be considered in situations where no information or no access to drive-by-wire is available for an existing vehicle:

■ Use a proprietary development kit, such as the ADAS Kit from Dataspeed Inc., which can be used to control selected versions of Ford and Lincoln vehicles [3].

■ Implement your own programmable interface for selected vehicles based on Open Source Car Control (OSCC) [11]. At the time of writing, the project only supports one specific Kia model. OSCC will be discussed in Chapter 5.

■ Retrofit the existing vehicle by replacing all the manual actuators with programmable ones, as shown in Figure 2.21.

Figure 2.21: Example of a retrofitted drive-by-wire vehicle. (Reprinted with permission from Rolling Inspiration. ©2018 Charmont Media Global)

2.4   In-vehicle networks

If sensors are the eyes and ears of an SDV and the computing platform is its brain, then the in-vehicle network is its central nervous system. Various types of communication networks can be used to connect hardware within the vehicle, and the choice depends on which sensors and vehicle platform controllers you are using. With so many types of sensors available nowadays, it is likely that the vehicle's computing platform will need to support more than one network system, for example *Controller Area Network (CAN)* and *Ethernet*.

CAN is the most common type of automotive network bus architecture. It has been employed since the late 1980s to support reliable message exchange between multiple vehicle controllers. Each CAN message contains an identifier and a payload of up to 8 bytes of data, as well as other information. The CAN message ID must be unique on each bus because it dictates how the data payload should be interpreted by the bus participants. CAN data transmission is limited to 1 Mbps. However, the improved CAN FD (flexible data-rate) development allows data to be transmitted faster than 1 Mbps and increases the maximum data payload to 64 bytes [4].

Table 2.1   Comparison of vehicle network systems

Name	Max. Bandwidth	Max. Payload	Real-Time	Cost
CAN	1 Mbps	8 bytes	no	low
Ethernet	1 Gbps	1500/9000[a] bytes	no	low
LIN	20 kbps	8 bytes	no	low
FlexRay	10 Mbps	254 bytes	yes	medium
MOST	150 Mbps	1014[b]/3072[c] bytes	no	medium

[a] Jumbo Frames
[b] MOST25
[c] MOST150

Source: Adapted from "An automotive Side-View system based on Ethernet and IP", by Alexander Camek, Christian Buckl, Pedro Sebastiao Correia, Alois Knoll, 2012, IEEE 26th International Conference on Advanced Information Networking and Applications Workshops (WAINA), p. 242.

Ethernet has been the de facto standard for LAN environments in the IT domain for over two decades, but has only recently begun to be adopted in series production vehicles due to technical issues such as radio frequency noise and latency [8]. These are critical to the automotive domain—though less relevant in the IT domain—and have only been solved recently. Ethernet enables high speed data transmission with larger data payloads than other vehicle communication networks. It also ensures efficient bandwidth use by offering a choice of broadcasting to all recipients, multicasting to a group of recipients, or sending to a single recipient. Multiple messages with short payloads can also be combined into one message to further reduce bandwidth utilization.

Table 2.1 summarizes the key differences between CAN, Ethernet and other automotive vehicle networks according to Camek et al. [2].

2.5   Summary

In this chapter, we discussed the major hardware components needed to transform a vehicle into an SDV: the various types of sensors, the computing platform, the actuator interface, and the in-vehicle network.

As we saw, there are many types of sensors available for use in SDVs. Exteroceptive sensors are designed to give the vehicle a view of its surroundings, acting as its 'eyes' and 'ears'. Active exteroceptive sensors work by transmitting energy and recording how long it takes to return. Radars and lidars both transmit electromagnetic energy. Radars measure the distance and direction to objects in their field of view using radio waves. Lidar produces 3D images of the surroundings in the form of point clouds using low-power lasers. Ultrasound sensors transmit sound energy and measure the distance and location of objects in the immediate vicinity of the vehicle. By contrast, passive exteroceptive sensors, such as cameras or GNSS receivers, passively record the energy they receive from the environment. Cameras are able to create detailed semantic maps of objects surrounding the vehicle. GNSS allows the vehicle to locate itself to within a few meters anywhere in the world (but only as long as it can 'see' sufficient satellites).

Proprioceptive sensors measure the state of the vehicle relative to some reference frame. IMUs bundle together a number of such sensors into one unit. Typically an IMU will measure 9 degrees-of-freedom (DoF) using triple-axis gyroscopes, triple-axis accelerometers, and triple-axis magnetometers. Finally, odometers are used to measure the speed and distance travelled for each wheel in the vehicle. As we will see in Chapter 3, IMUs and odometers enable a vehicle to perform dead-reckoning calculations for its position.

At the heart of any SDV is a high-performance computing platform. As we saw, these are in many ways similar to home PCs, but typically they employ specialized GPUs to perform the complex calculations required for SDVs (these calculations will be discussed in Chapter 3). Due to the safety-critical nature of SDV design, these platforms must support real-time requirements. The platforms also have to be able to operate in extreme conditions, and should draw the minimum-possible power.

We looked at how programmable actuators and by-wire controls are used to transform the commands issued by the computing platform into longitudinal (acceleration and braking) or lateral (steering) commands for the relevant ECUs. These commands are issued via an abstraction layer called the actuator interface. Due to the proprietary nature of many ECUs and actuators in modern vehicles, it may be necessary to retrofit a vehicle with actuators that you are able to control.

Finally, we discussed the in-vehicle network, an essential component of any modern vehicle and particularly vital for an SDV. We looked in detail at CAN and Ethernet, the two widely used networks in modern vehicles. We also looked at the relative merits of a number of different network types.

In the next chapter, we will look at how the data from the various sensors can be combined to allow an SDV to accurately perceive the environment around it. This is essential for knowing where it actually is (and how to get to its destination), for understanding the road conditions around it, and to avoid the risk of collision with other objects in the environment.

References

[1] Tanveer Abbas, Muhammad Arif, and Waqas Ahmed. Measurement and correction of systematic odometry errors caused by kinematics imperfections in mobile robots. In *SICE-ICASE, 2006. International Joint Conference*, pages 2073–2078. IEEE, 2006.

[2] Alexander Camek, Christian Buckl, Pedro Sebastiao Correia, and Alois Knoll. An automotive side-view system based on ethernet and ip. In *Advanced Information Networking and Applications Workshops (WAINA), 2012 26th International Conference on*, pages 238–243. IEEE, 2012.

[3] Dataspeed. Robot mobility base | adas kit | vehicle power distribution. http://dataspeedinc.com/what-we-make/. [Online; accessed 17-Aug-2018].

[4] Harald Eisele. What can fd offer for automotive networking. In *14. Internationales Stuttgarter Symposium*, pages 1237–1254. Springer, 2014.

[5] Heinrich Gotzig and Georg Otto Geduld. *LIDAR-Sensorik*, pages 317–334. Springer Fachmedien Wiesbaden, Wiesbaden, 2015.

[6] K. Hamada, Z. Hu, M. Fan, and H. Chen. Surround view based parking lot detection and tracking. In *2015 IEEE Intelligent Vehicles Symposium (IV)*, pages 1106–1111, June 2015.

[7] Hella. Wheel speed sensors in motor vehicles. function, diagnosis, troubleshooting. http://www.hella.com/ePaper/Sensoren/Raddrehzahlsensoren_EN/document.pdf. [Online; accessed 17-Aug-2018].

[8] Ixia. Automotive ethernet: An overview. `https://support.ixiacom.com/sites/default/files/resources/whitepaper/ixia-automotive-ethernet-primer-whitepaper_1.pdf`. [Online; accessed 17-Aug-2018].

[9] NASA. Global positioning system history. `https://www.nasa.gov/directorates/heo/scan/communications/policy/GPS_History.html`. [Online; accessed 17-Aug-2018].

[10] Society of Automotive Engineers. Vehicle Dynamics Committee. Vehicle dynamics terminology: Sae j670e : Report of vehicle dynamics committee approved july 1952 and last revised july 1976, 1978.

[11] PolySync. Open source car control. `https://github.com/PolySync/OSCC`, Oct 2018. [Online; accessed 17-Aug-2018].

[12] Konrad Reif. *Fahrerassistenzsysteme*, pages 321–367. Springer Fachmedien Wiesbaden, Wiesbaden, 2014.

[13] Arthur G Schmidt. Coriolis acceleration and conservation of angular momentum. *American Journal of Physics*, 54(8):755–757, 1986.

[14] Eyal Schwartz and Nizan Meitav. The sagnac effect: interference in a rotating frame of reference. *Physics Education*, 48(2):203, 2013.

[15] David H Sliney and J Mellerio. *Safety with lasers and other optical sources: a comprehensive handbook*. Springer Science and Business Media, 2013.

[16] US-EIA. Autonomous vehicles: Uncertainties and energy implications. `https://www.eia.gov/outlooks/aeo/pdf/AV.pdf`. [Online; accessed 17-Aug-2018].

[17] Mark Walden. Automotive radar - from early developments to self-driving cars. *ARMMS RF and Microwave Society*, 2015.

Chapter 3

Perception

The human brain is extremely good at perceiving the environment around it. Indeed, it's so good that we often take our ability for granted. When we cross a road, we look to see if there are cars approaching. If we see a car traveling towards us, we are able to rapidly estimate the speed at which it's approaching, and can decide whether it's safe to cross or not. Many things we do like this are pure reflex, and are based on prehistoric instincts for keeping us safe. When we look at a scene, our eyes just record patterns of light and color which are passed to the brain. Our brain then uses past experiences to interpret these patterns into an actual image. Because of the optics of the eye, we are actually seeing these images upside down, yet our brains are able to correctly interpret them, so that we can perform feats such as catching a ball. However, replicating this sort of ability in a self-driving vehicle is a challenging task.

As we saw in the previous chapter, SDVs rely on a range of different hardware sensors. But like our eyes, the raw data produced by these sensors is essentially meaningless. It is the job of the software to interpret this data and use that interpretation to build up a picture of the environment around the vehicle. More specifically, it is the perception and navigation software that performs this job of interpretation. If it sees a 1.5-meter tall, slender object slowly moving across the road in front, it might identify this as a pedestrian and pass that information to the control software. That software can then decide if the vehicle needs to take avoiding action or not.

The aim of the perception function is to achieve as complete and accurate understanding of the vehicle's environment as possible, in order to provide a basis for decision making in the subsequent navigation function. Perception involves answering the questions, 'where am I?' and 'what is around me?'. Reliable perception is crucial for ensuring smooth and safe operation of SDVs.

In general, perception in dynamic environments, i.e., environments with moving objects, can be decomposed into two major sub-functions: Simultaneous Localization And Mapping (SLAM) and Detection and Tracking of Moving Objects (DATMO). In this chapter, we will look in detail at how SDVs perceive the world around them.

3.1   Localization

Localization is the process of determining the vehicle's position and orientation based on a map. That map may be a global map for true self-driving vehicles traveling on the public roads, or it may be a more limited map for vehicles traveling within a bounded environment such as a factory.

Much research has been done into localization techniques for robots. However, there are some significant differences between these robots and SDVs. SDVs generally operate in environments that have already been accurately mapped. This helps simplify the localization problem. On the other hand, SDVs operate in a much more challenging and dynamic environment with fast moving objects and higher velocities.

There are two approaches to localization. *Local*, or *relative*, localization compares the current pose (or location) with the previous pose. *Global*, or *absolute*, localization uses external references to determine the current pose. These references might include satellites or known landmarks. Relative localization approaches are typically fast and require fewer resources compared to global approaches. However, they suffer from the effects of error or drift. More seriously, they can also fall victim to the *robot kidnapping problem*. This happens when they move to an arbitrary new location without correctly knowing their starting point. This might happen if the system reboots or loses state. In practice, this means usually both techniques are used in a complementary fashion. The relative location is used to track the current pose, but periodically, absolute localization is used to correct the results of any drift or to bootstrap the location after a system reset.

3.1.1 Localization based on GNSS

Global Navigation Satellite Systems (GNSSs) are a popular global localization technique as they offer a simple and inexpensive way for vehicles to localize. They use the principle of trilateration to determine the vehicle's absolute position anywhere in the world. However, this approach requires line-of-sight with at least three satellites, and therefore is not suitable for some operating environments where the satellites are obscured, e.g., indoors, urban canyon, tunnel, etc. Another disadvantage is the relatively low accuracy. The accuracy can be improved significantly using D-GPS or RTK base stations, but these are not available everywhere.

3.1.2 Localization based on wheel odometry

Wheel odometry localization is a relative localization approach that utilizes wheel sensors and a heading sensor. The localization is done by applying *dead-reckoning*, a simple technique used in ancient sea navigation. This estimates the vehicle position based on the projected direction and distance traveled relative to a known starting point. Because odometry-based localization does not need any external references, this approach works in any operating environments. As a relative localization method, this technique suffers from cumulative errors caused by wheel slip, uneven road surfaces, etc. Therefore, the localization result is typically only used in the short term to compensate for temporary unavailability of other localization techniques, as in the case of driving in a tunnel.

3.1.3 Localization based on INS

Similar to wheel odometry based localization, Inertial Navigation System (INS) localization is a relative localization technique that does not require any external references. INS-based localization is based on applying the dead-reckoning technique on motion and rotation measurements provided by an IMU (inertial measurement unit), which typically comprises of accelerometers, gyroscopes and magnetometers. Even though INS-based localization generally provides more accurate pose estimation than wheel odometry, it is still not resistant to accumulating errors, and so it needs to be corrected from time to time by other (absolute) localization techniques.

3.1.4 Localization with external references

Another way to achieve vehicle localization is by installing additional supporting devices or infrastructure within the operating environment.

The supporting infrastructure can take form of passive (non-transmitting) devices, such as magnets and visual markers, or active (transmitting) devices, such as Wi-Fi and Bluetooth beacons. Infrastructure-based localization is typically used for indoor environments where other localization approaches do not perform well. Depending on the technology and device arrangement used, robust and accurate positioning can be achieved. However, upgrading infrastructure is not always feasible, which makes this approach less suitable for operation in large areas. Zafari et al. [68] and Brena et al. [8] provide comparisons of common techniques and technologies using this approach.

3.1.5 Localization based on lidar

Lidar-based localization uses 'natural' landmarks, e.g., buildings, walls, trees, etc. that already exist in the operating environment. Since the localization does not require any special infrastructure to operate, this technique is more suitable for wider-scale operation, where installing additional infrastructure would be too expensive, or simply infeasible.

Lidars can be used for both local and global localization within a known map. The localization is usually done by performing *scan matching*. Scan matching is a technique that tries to find the geometric alignment of two scans, such that the scans optimally overlap each other. The resulting geometric alignment corresponds to the vehicle's translation and rotation that caused that change. By tracking the translation and rotation, the current pose can be estimated by incrementing the subsequent pose changes from the starting point. With regard to global localization, some scan matching techniques can also be applied to detect *loop closures*, i.e., whether the current location has been revisited due to the similarity of the current scan to a scan from a previous observation. Scan matching is also an effective method for the odometry pre-correction step in the map building process, which can lead to a substantial improvement of robustness and accuracy in large-scale environment mapping [28].

Scan matching

The most cited scan matching technique is undoubtedly the *Iterative Closest Point (ICP)*, originally introduced by Besl and McKay [5] and later adapted by Lu and Milios for localization applications [40]. ICP is an iterative algorithm that aims to minimize the point-to-point distances between two scans. The algorithm consists of three major steps:

1. For each point in the reference (or first) scan, find the correspondence by selecting the closest point or the nearest neighbor in the object (or second) scan.

2. Calculate the rigid body transformation that minimizes the mean square error of all correspondence pairs of the reference and the object scan.

3. Apply the transformation to the object scan and repeat until you achieve convergence.

Since its introduction in the 1990s, there have been many variants of ICP, as well as new proposals to improve the speed, stability, noise tolerance, etc. of the original ICP algorithm. *PLICP* is an ICP variant that uses a point-to-line distance metric instead of the point-to-point to enable faster convergence [10]. *Correlative Scan Matching*, also known as *Olson Scan Matching*, takes a probabilistic approach by formulating the problem as searching over the entire space of all plausible transformations (with respect to the motion commands or odometry), to find the rigid-body transformation that maximizes the probability of having observed the data [46].

Another class of approaches is based on feature-to-feature mapping, i.e., the scan points are extracted into a set of features, such as *Fast Laser Interest Region Transform (FLIRT)* [59], and the localization is thus performed by matching the currently observable features with a map or a database that contains the mapping of locations and their corresponding features. This approach can be particularly effective for environments with many locally distinguishable landmarks, and in general allows a more compact representation of the environment than its point-based counterparts.

Localization based on other range and bearing sensors, such as radars, has also been investigated. Vivet et al. utilized a slow rotating Frequency Modulated Continuous Wave (FMCW) radar for odometry and mapping [62]. Ward et al. achieved good localization results using low cost radars by applying extended Kalman filter-based localization in conjunction with ICP scan matching [63].

3.1.6 Localization based on cameras

Like range-finder based localization, camera, or visual-based, localization does not require additional infrastructure. The localization is done by exploiting the visual characteristics of the environment using mono, stereo or RGB-D (color and depth) cameras.

There are several approaches for localization using cameras. *Visual Odometry (VO)* localizes by estimating camera movement based on consecutive images. Similar to wheel odometry, the vehicle trajectory is obtained by incrementally updating the pose estimation since the starting point. In contrast, *Visual SLAM (VSLAM)* produces a globally consistent localization within the map, not only relative to the vehicle's starting point. VO can also be used in Visual SLAM to get a better motion estimate (compared to conventional wheel or INS-based odometry) [23].

VSLAM is a more difficult problem than VO and requires more computing resources, as it needs to solve *loop closure* detection and to keep track of all previous camera observations. Loop closure detection, or recognizing whether the current observation matches another observation in the past, is a general problem for SLAM techniques. When a loop closure is detected, a SLAM algorithm can correct the accumulated drift in both the map and the trajectory. This step is called *global bundle adjustment* (described later). Visual odometry sometimes performs bundle adjustment as well to create a more accurate estimate of the trajectory, but typically using a fixed number of recent observations (*windowed bundle adjustment*).

Visual-based localization methods generally follow two major approaches: the *feature-based* or the *appearance-based* approach. In the feature-based approach, the algorithm extracts key features such as edges and corners from the current observation and matches these with known features observed so far. The appearance, or direct, approach works directly using the image information, e.g., pixel intensities. The appearance-based approach is more suitable for low-textured or textureless environment; however, it is less robust to rapid brightness changes and image blurring [67].

One of the most popular feature-based visual localization algorithms is *ORB-SLAM2* [45], which works for mono, stereo and RGB-D cameras. ORB-SLAM2 uses *Oriented FAST and Rotated BRIEF (ORB)* features [49], that are basically a combination of a variant of the FAST keypoint detector and a rotation-aware BRIEF binary descriptor. As shown in Figure 3.1 ORB-SLAM2 uses three main threads that perform tracking, local mapping and loop closure detection in parallel. The tracking thread performs ORB feature detection and matching to the local map. The local mapping thread manages the local map and performs local bundle adjustment. Finally, the loop closure detection thread detects loop closing to avoid map duplication, and corrects any accumulated drifts. In order for the algorithm to detect loop closing or to relocalize, e.g., due to tracking failure, a database of ORB features based on *Discriminative Bags of Visual Words (DBoW2)* [24] is used and maintained in the place recognition module. A bag of visual words is a concept inspired from natural language processing. A visual word is an informative region made of a set of local features. Visual vocabulary is a collection of visual words, typically generated by clustering the features from a large set of image training data. Hence, in the bag of visual words concept, an image is represented by a histogram of the frequency of visual words obtained in that image, regardless of their spatial information.

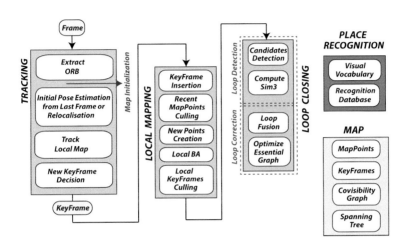

Figure 3.1: Overview of ORB-SLAM2 algorithm. (Adapted from "Orb-slam2: An open-source slam system for monocular, stereo, and rgb-d cameras", by Raul Mur-Artal, Juan D Tardós, 2017, IEEE Transactions on Robotics Volume 33 No. 5, p. 1255-1262. ©2017 IEEE)

The *Large-Scale Direct Monocular SLAM (LSD-SLAM)* [19] and *Stereo Large-Scale Direct SLAM (S-LSD-SLAM)* [20] are key algorithms for appearance-based visual localization. Both LSD and S-LSD SLAM algorithms operate directly on image intensities both for tracking and mapping, so they do not involve feature detection and matching as is the case with ORB-SLAM2. LSD SLAM (and its stereo counterpart) is basically a graph-based SLAM method that uses keyframes as pose-graphs. Each keyframe contains an estimated semi-dense depth map. The term 'semi-dense' here implies that not all image pixels are used, but only those pixels with a sufficiently large intensity gradient. As shown in Figure 3.2, the algorithm consists of three major tasks: tracking, depth map estimation, and map optimization. In the tracking step, the pose-graph constraints are determined by estimating the rigid body transformation using a direct image alignment method. In the depth map estimation step, either a refinement of the current depth map or a creation of a new depth map using the current frame as a new keyframe is performed. The latter is performed in case the camera moves too far away from the existing map. Finally, the map optimization step computes the optimal graph configuration, such that the errors imposed by the constraints are minimized. This can be solved by any generic graph-based SLAM back-end frameworks, such as the *General (Hyper) Graph Optimization (g2o)* [32] or *Sparse Bundle Adjustment (sSBA)* [30]. Graph-based SLAM is discussed in more detail in the last part of Section 3.3.

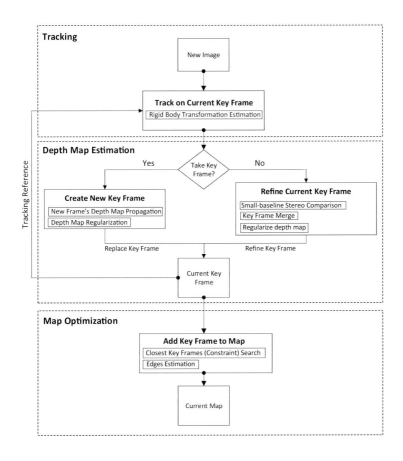

Figure 3.2: Overview of LSD-SLAM algorithm. (Adapted from "LSD-SLAM: Large-Scale Direct Monocular SLAM", by Jakob Engel, Thomas Schöps, Daniel Cremers, 2014, Computer Vision - ECCV 2014. Lecture Notes in Computer Science, vol 8690. Springer, Cham)

3.1.7   *Localization based on multi-sensor data fusion*

In reality, SDVs use a combination of the above approaches to get the optimal result in every situation. When GNSS-based localization is not reliable, e.g., driving between high buildings in the city, localization needs to rely on other methods, such as visual and lidar-based odometry. The technique of combining data or results from various sensors is called multi-sensor data fusion. This will be covered in Section 3.5.

3.2   Mapping

The pure localization techniques in the previous section work on the assumption that a highly precise and accurate map is already available beforehand. In reality, however, such high-definition maps are generally not publicly available and need to be generated most of the time. In this section, we will discuss the different types of maps commonly used for SDV. The choice of model or map type depends on several factors, including the type of sensors used, the memory and processing power of the computing platform, the localization algorithm used, etc.

3.2.1   *Occupancy grid maps*

Occupancy grid maps are undoubtedly the most popular type of maps in robotics and SDVs. A grid map discretizes the environment into a set (or grid) of square cells (or cube cells in 3D maps). Each cell in the grid map contains the occupancy probability of either occupied or free. Due to their generic representation, occupancy grid maps are also a popular map choice for multi-sensor data fusion.

3.2.2   Feature maps

Feature, or *landmark maps*, contain distinctive physical elements, e.g., trees, and their locations in the environment. Compared to grid maps, maps based on features have more compact representation due to their higher abstraction level, and are more robust to small variations in sensor observations. On the other hand, choosing the right features that work best for the specific SDV working environment might be challenging. In addition to that, performing feature extraction and matching online adds computational overhead. Figure 3.3 shows an example of a feature map taken from Victoria Park Sydney, where each feature is highlighted as a circle.

Figure 3.3: Example of a feature map. (Reprinted with permission from "Selective Submap Joining SLAM for Autonomous Vehicles", by Josep Aulinas, 2011, Doctoral dissertation, University of Girona, Spain)

3.2.3   Relational maps

A *relational map* defines the relationship between the elements of the environment. This is in contrast to the above map types, which work on the basis of the spatial information in the environment. One popular example of a relational map is the *pose-constraint map* as shown in Figure 3.4, mostly used in graph-based Simultaneous Localization and Mapping (SLAM). In a pose-constraint map, the elements in the map are vehicle poses, i.e., location and heading, built incrementally and represented using a graph. Elements in the map (or nodes in the graph) are connected to each other using edges, which represent the spatial constraints between the poses, typically based on odometry measurements.

Figure 3.4: Example of a pose-constraint map. (Adapted from the Victoria Park dataset in "Simultaneous localization and map building: Test case for outdoor applications", by Jose Guivant and Eduardo Nebot, 2002, IEEE Int. Conference on Robotics and Automation)

3.2.4   Other types of maps

There are several other map representations besides the ones mentioned above. Whereas a grid map stores the occupancy information of all cells, a *point-based map* is a more compact map type that only contains the occupied information, such as a set of 3D point clouds of solid objects detected by lidar sensors. In contrast, a *free-space map* is another memory-optimized map type that holds only the free space information. The free space can be represented by geometrical shapes, e.g., trapezoids, cones, etc. or Voronoi graphs. Another popular map type is a *line map*, which uses a set of lines to represent the environment. A comparison of common map types with regard to compactness, required computing speed, level of detail, and other aspects is provided in [52].

3.3   SLAM

As we saw in the previous section, building an accurate map is only possible when the exact pose of the vehicle is known. In reality, determining the exact pose of the vehicle is a difficult task due to sensor measurement errors. Simultaneous Localization and Mapping (SLAM) tries to solve this classic chicken-and-egg problem. SLAM is a technique that seeks to build an accurate map of the environment while simultaneously localizing the vehicle within that map. As shown in Figure 3.5, the problem is that both the observed map features and the vehicle's reported location suffer from errors, and these errors increase the further the vehicle travels from its last known location. However, thanks to loop closure, once the whole route has been driven in a loop more than once, SLAM can generate a consistent and accurate map. The resulting map can then be used to accurately localize the vehicle anywhere within that map by performing the same SLAM algorithm in localization mode only (without mapping).

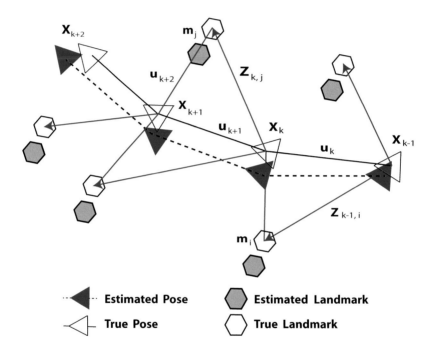

Figure 3.5: SLAM problem. Due to measurement errors, drifts, etc. the true locations of the vehicle are never known. The estimated pose x_{k+1} is calculated by applying the vehicle control u_{k+1} to the the previous estimated pose x_k. The term $z_{k,j}$ denotes the measured distance between the vehicle at pose x_k and the observed landmarks m_j. (Adapted from "Simultaneous Localisation and Mapping (SLAM) : Part I The Essential Algorithms", by Hugh F. Durrant-Whyte and Tim Bailey, 2006, IEEE Robotics and Automation Magazine Volume 13 Number 2, p. 99-108)

In classical robotics literature, the SLAM problem is sometimes divided into *full* and *online* SLAM problems. Full SLAM aims at estimating the entire path and map, whereas online SLAM only estimates the most recent pose and map. Given $X_T = \{x_0, x_1, \ldots, x_t\}$ the path or sequence of travelled locations x_0 as the starting point, m the map of the environment, $Z_T = \{z_1, z_2, \ldots, z_t\}$ the sequence of measurements, and $U_T = \{u_1, u_2, \ldots, u_t\}$ the sequence of control inputs, the full SLAM problem can be mathematically formulated as Equation 3.1 follows:

$$p(X_T, m | Z_T, U_T) = p(x_{0:t}, m | z_{1:t}, u_{1:t}) \tag{3.1}$$

whereas the online SLAM is formulated as Equation 3.2 as follows:

$$p(x_t, m | z_{1:t}, u_{1:t}) = \int_{x_0} \int_{x_1} \cdots \int_{x_{t-1}} p(x_{0:t}, m | z_{1:t}, u_{1:t}) \, dx_{t-1} \, \cdots \, dx_1 \, dx_0 \tag{3.2}$$

Since the SLAM problem was first described in the mid-1980s, several algorithms have been proposed that solve the problem [66]. In general, SLAM algorithms can be classified into two major approaches: the *filtering approach*, and the *optimization approach*.

3.3.1 Filtering approach

The filtering approach refers to a technique that estimates the current unknown state based on past observations. Hence, it iteratively improves its internal belief by incorporating new observations as they are gathered. There are two significant variants of this approach: the Kalman filter and the particle filter.

3.3.1.1 Kalman filter

Kalman filters are a class of Bayesian filters that make the assumption that all noise in the system is Gaussian. As shown in Figure 3.6, the Kalman filter is a recursive method with two major steps: the prediction step and the update step. The prediction step uses the most recent state estimate $\hat{x}_{k-1|k-1}$ and the error covariance estimate $P_{k-1|k-1}$ from the previous iteration (or initial estimate) to compute the predicted state estimate $\hat{x}_{k|k-1}$ and predicted error covariance $P_{k|k-1}$. The update step involves correcting the predicted state estimate calculated in the previous step by taking the most recent measurement y_k into account to generate an updated state estimate $\hat{x}_{k|k}$ and updated error covariance estimate $P_{k|k}$. The *Bayesian filtering approach* is a probabilistic approach that uses the recursive Bayesian inference framework to estimate the unknown probability distribution, i.e., the estimated states. Two of the most popular

SLAM algorithms using this approach are the Extended Kalman Filter (EKF-SLAM) and the Unscented Kalman Filter (UKF-SLAM).

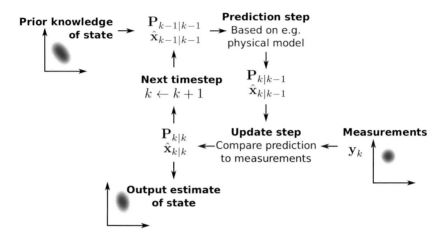

Figure 3.6: Kalman filter algorithm. (©Petteri Aimonen / Wikimedia Commons / CC-Zero-1.0)

The *Extended Kalman Filter (EKF)* and *Unscented Kalman Filter (UKF)* can be used to solve systems that involve non-linear motion and/or measurement models. This makes them ideal for use in SDVs, since common sensors such as cameras, lidars, ultrasound, and radars use a polar coordinate system (angle and distance), which introduces non-linear terms when converted to Cartesian coordinates (x, y, z). Also, motion models typically involve translation and rotation, which might be non-linear.

While linear KFs are proven to always produce an optimal solution for linear systems, EKF and UKF can only give an approximation to the non-linear probability distribution of the estimated state. EKF approximates the non-linear probability distribution by employing linearization based on the Taylor first order expansion. By contrast, UKF is based on the *Unscented Transformation (UT)* that uses a set of specially chosen weighted points, i.e., sigma points, to represent the unknown probability distribution. The number of points needed to preserve the covariance and mean of distribution is defined as $2L + 1$, where L is the number of the dimension. So for a two-dimensional probability distribution, five points are needed: one mean and two points on each side of the mean. The Kalman filter update equations apply those sigma-points directly to the non-linear functions. Thus, it is no longer necessary to compute the derivation of the non-linear terms, i.e., the Jacobian matrix, as in the EKF case. On some models, this derivation might be difficult to obtain and may not even exist in closed form.

EKF-SLAM and UKF-SLAM are simply EKF and UKF applied to solve the SLAM problem, respectively. EKF-SLAM was the first solution proposed for the online SLAM problem, first described by Smith, Self and Cheeseman [55]. Due to their structures, both methods are particularly useful for working with landmark-based maps, i.e., maps that maintain landmark positions, and take almost the same computational time. However, both methods are known to be less useful for large-scale maps since the complexity increases significantly as the number of landmarks increases. Hence EKF-SLAM and UKF-SLAM are typically only used for SDVs operating in small-scale environments, such as confined spaces or private roads with distinguishable landmarks.

3.3.1.2 *Particle filter*

The Kalman filter approach in the previous section relies on the strong assumption that the unknown probabilistic distribution is a Gaussian one. Even though this assumption may work well in many use cases, the Gaussian distribution is too simple for some other scenarios. Another limitation of the Kalman filter approach is that it only allows one hypothesis at a time for the current state (map and pose), due to its unimodal probabilistic distribution model. Imagine two different locations in a map with similar sensor measurement signatures. Because only a single hypothesis is possible at any time, the filter has to pick one of the two candidate locations. If the filter chooses wrongly, all subsequent observations will be inconsistent with its previous beliefs from the history. Hence, the filter gets 'stuck' in its wrong belief, and will never converge, unless it finds a way to recover itself from this situation or gets reinitialized.

The Particle Filter (PF) approach overcomes these limitations by lifting the Gaussian noise assumption. In other words, the approach makes no assumption about the unknown probabilistic distribution and is therefore more suitable for complex non-linear, non-Gaussian models. The approach is based on a technique called *Sequential Monte Carlo (SMC)*, which approximates an unknown probabilistic distribution as the sum of weighted samples or 'particles', drawn randomly from a *proposal distribution* using the *importance sampling* method [29].

Note that, because the target distribution we want to approximate is unknown, we cannot draw any samples from it. However, we can use a statistical trick called *importance sampling* to generate samples *indirectly*, i.e., by drawing samples from a different probability distribution (known as the *proposal distribution*) and assigning weights (called the *importance weight*) to all the samples such that their weighted density is proportional to the target distribution. Figure 3.7 illustrates the principle of importance sampling.

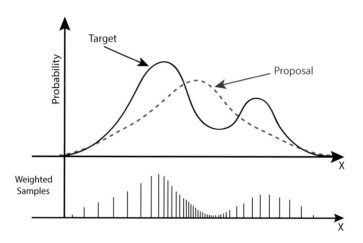

Figure 3.7: Importance sampling. The unknown target distribution $f(x)$ **is approximated by drawing samples from a proposal distribution** $g(x)$ **with importance sampling weight** $w(x) = f(x)/g(x)$**. (Adapted from "FastSLAM. A scalable method for the simultaneous localization and mapping problem in robotics", by Michael Montemerlo, Sebastian Thrun, 2007, Springer Tracts in Advanced Robotics, p. 37. ©2007 Springer Berlin Heidelberg)**

In the context of SLAM, each particle in the particle filter contains a concrete hypothesis of the map and vehicle pose. In each iteration, the estimated map and pose of each particle is updated according to the vehicle's motion model and sensor measurements. Also, the weight of each particle is reassigned based on the observation likelihood. However, it is well known that PF suffers from the *curse of dimensionality*, because the number of particles needed to ensure consistent density with the target distribution grows exponentially with the system dimension [3].

The obvious way to mitigate this problem is to reduce the size of the state-space. A standard approach to do this is by applying the Rao-Blackwellization technique to the PF, resulting in the *Rao-Blackwellized Particle Filter (RBPF)*. The idea is based upon the observation that if we know the trajectory, we can build the map. Due to the strong dependency between map and pose estimates, it is no longer necessary to sample both distributions, which makes RBPF a much more efficient method than the standard PF.

RBPF-SLAM decomposes the SLAM problem into two state sub-spaces, namely the trajectory sub-space and the map sub-space conditioned by the trajectory, which is formulated as follows:

$$p(x_{1:t}, m|z_{1:t}, u_{0:t-1}) = p(x_{1:t}|z_{1:t}, u_{0:t-1}) \cdot p(m|x_{1:t}, z_{1:t}) \quad (3.3)$$

It can be proved that the posterior over the map space (the second component of the posterior) in Equation 3.3 can be factorized as follows [58]:

$$p(m|x_{1:t}, z_{1:t}) = p(x_{1:t}|z_{1:t}, u_{0:t-1}) \cdot p(m|x_{1:t}, z_{1:t}) \quad (3.4)$$

By applying the factorized posterior to Equation 3.4, the full posterior becomes:

$$p(x_{1:t}, m|z_{1:t}, u_{0:t-1}) = p(x_{1:t}|z_{1:t}, u_{0:t-1}) \prod_{n=1}^{N} p(m_n|x_{1:t}, z_{1:t}) \quad (3.5)$$

Note that the first term of the posterior, $p(x_{1:t}|z_{1:t}, u_{0:t-1})$, is basically a localization problem based on observations Z and motion controls U, which can be solved using a standard PF. The map estimation in the second term can be computed efficiently using a Kalman filter, e.g., EKF as in the popular FastSLAM algorithm [43]. In FastSLAM, each particle contains a trajectory estimate and a set of low-dimensional EKFs that individually estimate the location of each landmark in the map.

One common challenge with PF is the so-called *particle degeneracy* problem. This is where almost all particles have negligible weights in the long term. There are two major approaches to prevent degeneracy, namely resampling and better proposal distribution.

The basic idea of resampling is to keep the most likely particles and replace the unlikely ones, so as to reduce the number of low-weight particles. However, a good resampling strategy is crucial, lest the PF falls into another problem known as *particle impoverishment*. The particle impoverishment problem is a situation where the filter loses its diversity, i.e., it starts to rely on a small number of particles with significant weight while the majority of other particles with small weights are abandoned. This situation certainly reduces the multiple hypotheses advantages of PFs, and the abandoned particles might be important ones. An overview of common resampling algorithms, as well as the comparison between them, is provided in [16]. Part of a good resampling strategy is to know when (how frequently) to resample, because every resampling step increases variance. A general rule is to resample whenever the *Effective Sample Size (ESS)* is below a certain threshold [34]:

$$N_{ESS} = \frac{N}{1 + N^2 Var(w_{k|k}^i)} \tag{3.6}$$

where N is the number of particles and $Var(w_{k|k}^i))$ the variance of the weights of all the particles. Equation 3.6 is sometimes simplified using the following approximation [27]:

$$\hat{N}_{ESS} \approx \frac{1}{\sum_{i=1}^{N}(w_{k|k}^i)^2} \tag{3.7}$$

Another approach to improve PF performance is through using a better proposal distribution. A proposal distribution is described as optimal if it minimizes the conditional variance of the importance weights [17]. Intuitively, the better the proposal distribution, the closer it matches the true posterior and the more often a new sample drawn from it 'agrees' with the observations. One way to do this is to incorporate recent observations in the proposal distribution, an approach taken by FastSLAM 2.0 (the improved version of FastSLAM). The optimal proposal distribution not only requires fewer particles to achieve the same performance as the standard approach (due to the reduction in variance), but it also makes the algorithm more robust to large motion uncertainty [43].

3.3.2 Optimization approach

In essence, the filtering approach works by summarizing information of past observations as new measurements are gathered sequentially and propagating this information through time. The information here involves the joint probability distributions of all features in the map (how the features are interconnected) and the current pose. All poses other than the current one are marginalized out. By contrast, the optimization method works using a *smoothing principle*. This means that all poses and measurements from the start until the current observation are used to find the most likely overall trajectory, defined as the one most consistent with the whole observation set.

Because all past observations are taken into account, the optimization approach is actually a solution to the full SLAM problem. However, this does not mean that the optimization approach can only be done using an offline batch operation. As we will see in the following section, some optimization approaches are also suitable for solving the online SLAM problem too. In fact, the optimization approach has been dominating state-of-the-art SLAM algorithms and is a key enabler for building large-scale maps [56].

Methods based on the optimization approach are typically made up of two processes: front-end and back-end, as shown in Figure 3.8. The *front-end* is the sensor-dependent process, which is responsible for extracting the relevant features from the sensor data, and performing data association. The local or short-term data association is performed by tracking the features between two consecutive measurements, usually to 'correct the odometry', i.e., to get a better estimation of the vehicle movement. This step is important in order for the algorithm to generate a consistent map. A global, or long-term, data association is performed to detect loop closing, i.e., whether the vehicle revisits a place in the observation history. The *back-end* process is responsible for finding the optimal configuration that is maximally consistent with all observations. Formally, this task is called solving the *Maximum A Posteriori (MAP)* estimation problem, which is defined as follows:

$$\mathbf{X}^* = \arg\max_{X} p(X|Z) \tag{3.8}$$

where $\mathbf{X}^*$ is the optimal configuration of random variables, X, that maximizes $p(X|Z)$, the *belief* over measurement Z. Applying Bayes' theorem, Equation 3.8 can be rewritten as:

$$\mathbf{X}^* = \arg\max_{X} \frac{p(Z|X)p(X)}{p(Z)} \propto \arg\max_{X} p(Z|X)p(X) \tag{3.9}$$

where $p(Z|X)$ is the probability of measurements Z given the configuration X, and $p(X)$ is the prior probability over X. Note that in the case of an unknown prior probability or where the prior is assumed to have a uniform distribution, the term $p(X)$ in Equation 3.9 becomes a constant and the MAP estimation is reduced to the *Maximum Likelihood Estimation (MLE)* problem.

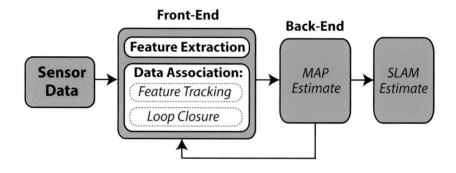

Figure 3.8: SLAM front-end and back-end processes.

We will now examine two major variants of the optimization approaches: Graph-based SLAM and Bundle Adjustment (BA).

3.3.2.1 Graph-based SLAM

Graph-based SLAM addresses the SLAM problem by constructing a pose-constraint graph and finding the configuration that is maximally consistent with the graph. Each node in the graph represents a vehicle pose, and is connected to another node by an edge that represents the spatial constraints (translation and rotation) between the poses. The graph creation and data association are done by the sensor-specific front-end component. The back-end component solves the maximum likelihood estimation by employing some non-linear square methods, such as the *Gauss-Newton* or *Levenberg-Marquardt* algorithms.

Let us use a simple example to show how graph-based SLAM works. Starting from an arbitrary initial position, we give the vehicle a command to accelerate, e.g., from zero to around 10 km/h in a second, in a straight direction, i.e., 0 degree steering angle. In an ideal world, the vehicle should now be located around 2.78 meters away from the starting position. However, due to wheel friction, energy loss, etc., our odometry sensor only reads less than this value, say 2 meters. Let us assume our vehicle also has a front-looking camera that performs visual odometry. The visual odometry tends to be more accurate than our wheel odometry sensor, so we take the corrected values from the visual odometry instead. It turns out that according to visual odometry, our vehicle actually moved 2.5 meters with a 3-degree deviation from the initial position. The odometry correction from two consecutive observations is the short-term data association part of the SLAM front-end. Now our graph has two nodes, one for the initial pose and another for the current pose. The nodes are connected by an arc that describes the translation and rotation values resulting from our short-term data association. The long-term data association part detects loop-closing, and determines whether the current observation matches one previously in the history. In order to be able to do this, each node stores some information about the environment that the sensor perceived in that pose. Ideally, the information should be able to uniquely identify a previous location in the map without consuming significant CPU and memory resources for extraction, comparison and storage. Depending on the sensor technology used, the information used for the data association might be a Bag of Visual Words (BoVW), landmark coordinates, etc.

Each node in the graph is also called a *keyframe*, and it represents a snapshot of the current state in a continuously traveling vehicle. How often a keyframe is created in the graph is typically a performance/accuracy trade-off, and is determined through experiment. If keyframes are created too frequently, more computation and memory resources are needed for the graph, and for the front-end and back-end processes. On the other hand, less frequent keyframes consume less resources, but the algorithm might risk missing an important loop-closing situation. The configuration also depends on the driving scenario. For the same distance traveled, e.g., 10m, significantly more relevant information will be gathered in an urban scenario compared with driving on a highway.

The back-end process performs the optimization of the underlying graph by finding a configuration of all the nodes in the graph that maximizes the likelihood of all observations. Recall that, in case of unknown prior probability or assuming uniform prior probability distribution, the MAP problem is equivalent to MLE:

$$X^* = \arg\max_{X} p(Z|X) \tag{3.10}$$

Assuming each observation is independent, the overall likelihood of all observations as the product of the individual likelihoods is defined in Equation 3.11 as follows:

$$L(Z|X) = \prod_{i=0}^{n} L(Z|x_i) \qquad (3.11)$$

where $L(Z|x_i)$ is the likelihood of measurement Z given the individual configuration x_i.

A common practice is to rewrite the above equation as a log-likelihood function, so that the overall likelihood now becomes an additive function as described in Equation 3.12:

$$\log L(Z|X) = \log \prod_{i=0}^{n} L(Z|x_i) = \sum_{i=0}^{n} \log L(Z|x_i) = \sum_{i=0}^{n} l(Z|x_i) \qquad (3.12)$$

where $l(Z|x_i)$ is equal to $\log L(Z|X_i)$.

It turns out that finding the optimal configuration X^* that has the maximum likelihood of all observations is equivalent to finding a configuration where the sum of differences between all expected and real measurements over all nodes is at a minimum. This deviation between the real and expected measurements is called the error (or cost) function, which is formulated as follows:

$$e_{ij}(x_i, x_j) = z_{ij} - \hat{z}_{ij}(x_i, x_j) \qquad (3.13)$$

where z_{ij} is the real measurement from node x_i to x_j and $\hat{z}_{ij}$ the expected measurement between the two nodes, respectively. By modeling the uncertainty of measurement as Ω_{ij} and C as the set of constraint pairs in all observation Z (see Figure 3.9), we seek to find the optimal configuration X^* that minimizes the negative log-likelihood function $F(x)$ as defined in Equation 3.14:

$$X^* = \arg\min_{x} F(x) \qquad (3.14)$$

where:

$$F(x) = \sum_{<i,j>\in C} e_{ij}^T \, \Omega_{ij} \, e_{ij} \qquad (3.15)$$

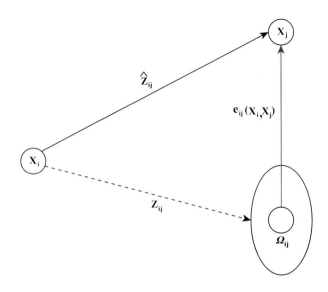

Figure 3.9: Graph-based SLAM cost function model, where z_{ij} denotes the real measurement from node x_i to x_j, $\hat{z}_{ij}$ the expected measurement between the two nodes, Ω_{ij} the uncertainty of measurement and $e_{ij}(x_i, x_j)$ the deviation between the real and expected measurements as the error (cost) function.

Equations 3.14 and 3.15 form what is commonly known as the least-squares minimization over the *objective (cost) function* $F(x)$. There are various approaches for solving this problem in the literature. However, SLAM, as with other non-linear least-squares problems, generally does not have any closed-form solutions [25]. Therefore, solving the problem typically requires an algorithm that starts with an initial value, either randomly selected, guessed or heuristics-based, and iteratively minimizes the cost function until convergence. Some popular standard solvers are the *Gradient Descent (GD)*, *Gauss-Newton (GN)* and *Levenberg-Marquardt (LM)* algorithms. GD iterates with steps proportional to the negative of the gradient, until the local minimum is reached. Due to using the gradient, i.e., the first-order derivation of the cost function, GD is sometimes also called the first-order optimization approach. GN 'linearizes' or approximates the cost function using first-order Taylor expansion at each iteration, and calculates a new step by solving the resulting linear system. LM is a combination of GD and GN. If the parameters at the current iteration are far from the optimal value, LM takes a larger step (or *damping factor*) and acts more like GD. On the other hand, if the current parameters are close to the optimal value, LM takes a smaller step and behaves more like GN.

Applying the above standard non-linear least-square solvers directly to the SLAM problem may lead to a sub-optimal result because the parameters, i.e., the configurations X, are assumed to be in Euclidean space [26]. Therefore, modern SLAM back-ends typically perform *on-manifold* least-squares optimization, as the space of the rotational parameter is not Euclidian. A manifold is a topological space that locally resembles Euclidean space, but globally might not be Euclidean [33]. On-manifold optimization has basically the same structure as the Euclidian counterpart. For each iteration, a new step is computed in the local Euclidian approximation space. The accumulated increments are projected in the global non-Euclidian space and the process is repeated until convergence [26].

Some open-source implementations of graph-based SLAM are available on the Internet. A list of some popular ones is provided in [56].

3.3.2.2   Bundle adjustment

Bundle Adjustment (BA) is a visual reconstruction technique that aims to jointly optimize the 3D structure and the parameters (pose and/or calibration), from which the model is viewed [60]. The *bundle adjustment* method was originally developed for aerial mapping, where it is used to optimally adjust the bundle of *light rays* that emanate from each image feature, so that they all converge around one single point, i.e., the center of the camera focal plane. As shown in Figure 3.10, the method looks for the optimal adjustment of *bundled* light rays, such that the rays from each feature and reference landmark converge around the camera center.

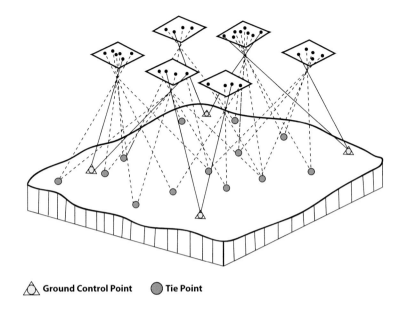

△ **Ground Control Point**   ⬤ **Tie Point**

Figure 3.10: Bundle adjustment in aerial mapping. The ground control points are fixed (usually physically marked) points on the Earth's surface with known ground coordinates. Tie points refer to identifiable features with unknown coordinates that can be used as reference points.

Similar to graph-based SLAM, BA is generally formulated as a nonlinear least-squares problem whose goal is to find the optimal configuration that minimizes a cost function. A commonly used cost function is the re-projection error, i.e., the difference between the observed feature location and the expected 2D projection of each corresponding 3D point onto each image plane, which is formulated in Equation 3.16 as follows [2]:

$$\arg\min_{a_j, b_i} \sum_{i=1}^{n} \sum_{j=1}^{m} d(P(a_j, b_i), x_{ij})^2 \tag{3.16}$$

where vector a_j represents the camera parameters, b_i all visible 3D points in image j, x_{ij} is the corresponding 3D feature point i on image j, $P(a_j, b_i)$ is the estimated projection of 3D point i on image j and $d(x, y)$ denotes the *Euclidian distance* between the image points represented by vectors x and y, as shown in Figure 3.11.

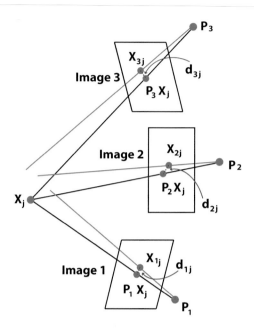

Figure 3.11: Total 2D reprojection error (d_{1j} + d_{2j} + d_{3j}) as a BA cost function. The reprojection error d_{ij} is defined as the Euclidian distance between the point X_j and the P_iX_j (the estimated projection of point X_j on image i).

Applying BA to the general SLAM problem, we can reformulate the problem as finding an optimal configuration that minimizes the difference of the observed landmark locations, and the expected projected location of each landmark due to the vehicle's movement. Thus, the cost function is the sum of landmark observation error and the odometry error, or the pose-to-landmark and the pose-to-pose error, respectively [22]. The set of pose-to-landmark constraints is basically a map, because it contains the locations of all observed landmarks. Note that, if we ignore the pose-to-landmark constraint, or marginalize out the map, BA reduces to the graph-based SLAM problem. Hence, graph-based SLAM can also be called a special instance of BA, which can be applied to scenarios that do not use any landmarks or features in the map [22].

Note that having formulated a non-linear least-squares problem, we can also solve BA by applying standard methods mentioned at the end of the graph-based SLAM section. Typically, the algorithm of choice to solve the BA problem is the Levenberg-Marquardt (LM) algorithm. However, due to the large number of unknown variables involved in the cost function, directly applying LM to the BA problem is computationally very expensive, which makes it less suitable for solving SLAM online. Therefore, some optimizations have been proposed to reduce the complexity. One such proposal, the *Local Bundle Adjustment (LBA)* or the *sliding time window approach*, is to perform the least-squares optimization only for the n most recent images and consider only the 2D reprojections in the last N frames [44]. Another approach is to use the *Dog Leg* minimizing algorithm, which can achieve the same quality as LM but in a much shorter computation time [38]. An overview of the current BA optimization approaches is provided in [37].

3.4 Object detection

One of the basic skills that an SDV needs is object detection. Not only is object detection essential for the SDV to drive safely, i.e., to prevent collisions/accidents, but it is also important to gain a proper understanding of the environment, so it is able to make the best possible decision for the current situation. As human drivers, we perform extensive simultaneous object detection tasks, sometimes unconsciously. We need to recognize not only other moving objects in the environment, such as cars, pedestrians, bicycles, but also static objects like the lane boundaries, traffic signs, traffic lights, and many other hazards. It is still very challenging for computers to replicate this ability today. However, some promising technologies, such as deep learning (which will be covered briefly later in this section and in more detail in Chapter 7), mean the gap is closing rapidly.

In computer vision literature, object detection is typically divided into the following sub-problems:

■ Object localization, i.e., determining the bounding box of detected objects.

■ Object classification, i.e., categorizing the detected objects into one of the pre-defined classes.

■ Semantic segmentation, i.e., partitioning the image into semantically meaningful parts and classifying each part into one of the pre-determined semantic regions.

Figures 3.12 and 3.13 illustrate the differences between the above subproblems.

Figure 3.12: Object localization and classification. (Adapted from "Cars driving on a rainy day" by Good Free Photos. ©2018 GoodFreePhotos.com / CC-Zero-1.0)

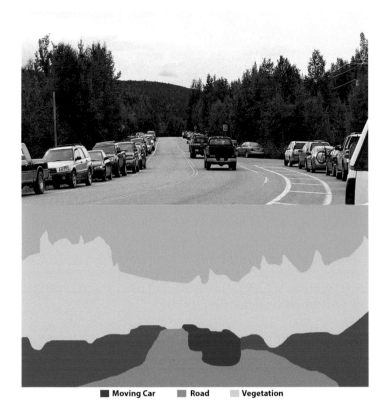

Moving Car   **Road**   **Vegetation**

Figure 3.13: Semantic segmentation. (Adapted from "Old Nenana High-way, Ester, Alaska" by User:RadioKAOS. ©RadioKAOS, https://commons.wikimedia.org/wiki/File:Old_Nenana_Highway,_Ester,_Alaska,_showing_cars_lining_the_road_during_Angry,_Young_and_Poor_Festival.jpg,   "Old Nenana Highway, Ester, Alaska, showing cars lining the road during Angry, Young and Poor Festival", Coloration, https://creativecommons.org/licenses/by-sa/3.0/legalcode)

Object detection has been actively researched in computer science since the mid-1960s. There have been many proposals to solve this problem. [21] provides a brief chronological outline of the popular approaches. In general, solving the object detection problem involves the following steps (see Figure 3.14):

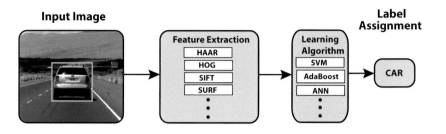

Figure 3.14: Generic object detection pipeline.

- Pre-processing
  The pre-processing step 'normalizes' the image, i.e., it performs some adjustments to the raw image, such that it matches the expected input for the subsequent feature extraction step. This might involve image rotating and resizing, intensity adjustment, and so on. The actual tasks to be performed are application-specific. Some approaches even skip the pre-processing step altogether.

- Feature extraction
  The feature extraction step removes unimportant or extraneous information from the image, and preserves only the relevant information (or features) for the classification. This transforms the image into another representation known as the feature map.

- Classification
  The final step matches the feature map with the reference feature maps that represent each of the pre-defined classes.

3.4.1 Feature extraction

The main challenge of object detection lies in the *feature engineering*, i.e., designing feature descriptors that make each class clearly distinguishable from every other. Imagine we want to make a simple classification for tables, i.e., we want the computer to correctly output 'table' if its input is a table image and 'not a table' otherwise. A very simplistic feature descriptor might be a function that states 'a table has four legs'. Based on this feature descriptor, we extract only the relevant information about leg-like parts of the image, and discard all other information from the image. In the classification step, the count of the leg-like image parts will be determined, and a decision is made based upon that information. Using a poor feature descriptor like this will certainly lead to poor object classification results, no matter how well the classifier in the last step performs. There are many tables that do not have four legs, such as one-legged bar tables. Equally there are many things in this world that have four legs which are not tables, such as dogs.

Fortunately, a number of generic feature descriptor algorithms have been defined and successfully applied to solve a wide range of object detection problems. Some of the most popular ones are described below.

3.4.1.1 *Histogram of oriented gradients*

The *Histogram of Oriented Gradients (HOG)* descriptor [12] uses the distribution (or histogram) of intensity gradients, or edge directions, as a representation of local object appearance and shape. As shown in Figure 3.15(b), the algorithm divides the image into small cells, and a histogram of gradient directions for each pixel in the cell is calculated. Typically, the contrast of local histograms is normalized using the average intensity value across a block or a set of connected cells, before all local histograms are then concatenated to form the final descriptor.

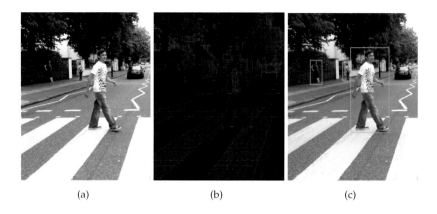

(a)   (b)   (c)

Figure 3.15: (a) Original image. (b) The corresponding HOG features. (c) The detected objects in bounding boxes. (Adapted from "Abbey Road Crossing", by Sparragus. ©Sparragus, https://commons.wikimedia.org/ wiki/File:Abbey_Road_Crossing.jpg, "Abbey Road Crossing", Coloration, https://creativecommons.org/licenses/by/2.0/legalcode)

3.4.1.2   Scale-invariant feature transform

Scale-Invariant Feature Transform (SIFT) [39] represents an image as a set of invariant keypoints, i.e., local features in the image that are invariant of rotation, translation, scale, illumination and other viewing conditions. For each keypoint, a vector of 128 numbers is calculated as its *fingerprint* using a histogram of gradient magnitudes and orientations as shown in Figure 3.16(b) and Figure 3.16(d). Finally, the fingerprint can be compared with a set of known fingerprints to determine if the object is the same as depicted in Figure 3.16(e).

Figure 3.16: Object matching (e) by comparing the SIFT features (b) and (d) calculated from the original images (a) and (c). (Adapted from "Glyptothek in München in 2013" by High Contrast and "München Glyptothek GS P1070326" by Georg Schelbert. © High Contrast, https://commons .wikimedia.org/wiki/File:Glyptothek_in_München_in_2013.jpg, "Glyptothek in München in 2013", Coloration, https://creativecommons.org/licenses/by/3.0/ de/legalcode). © Georg Schelbert, https://commons.wikimedia.org/wiki/ File:München_Glyptothek_GS_P1070326c.jpg, "München Glyptothek GS P1070326c", Coloration, https://creativecommons.org/licenses/by-sa/3.0/ legalcode)

3.4.1.3 Maximally stable extremal regions

Maximally Stable Extremal Regions (MSER) [42] is a blob detection method that works by detecting the property changes of a region (or a set of connected pixels) relative to its surroundings. As shown in Figure 3.17(a) and Figure 3.17(b), MSER describes an image as a set of regions that are maximally stable, or virtually unchanged, despite intensity changes. In other words, it seeks to find regions that remain visible across a wide range of brightnesses. The MSER region is usually described using an ellipsoid that is fitted to the actual shape. Compared to SIFT, MSER is faster and invariant to affine transformations, such as skewing [51].

(a)   (b)

Figure 3.17: (a) Original image. (b) The corresponding MSER features in bounding boxes. (Adapted from "Chevry-sous-le-Bignon" by François Goglins. ©François GOGLINS, https://commons.wikimedia.org/wiki/File:Chevry-sous-le-Bignon-FR-45-carrefour_D33_&_D146-e.jpg, Coloration, https://creative commons.org/licenses/by-sa/4.0/legalcode)

Open-source implementations of HOG, SIFT, MSER, as well as several other feature descriptor and extraction algorithms, can be found at the VLFeat [61] or OpenCV [47] project.

3.4.2 Classification

The final step of the object detection task is classifying the features extracted in the previous step into a set of pre-defined classes, such as 'car', 'pedestrian', 'truck', etc. Typically, the classification task is performed by a machine-learning classifier algorithm. Some of the widely used classifiers include support vector machine (SVM), random forest and artificial neural networks (ANNs).

3.4.2.1 Support vector machine

Support Vector Machine (SVM) [6] is one of the most popular and efficient algorithms for classification, which aims to find a separating hyperplane that optimally separates sets of different class labels. In most cases, it might be impossible to separate the classes using a merely linear function. However, the non-separable data might be linearly separable in a high-dimensional space, and an optimal separating hyperplane can be determined. Thus, with the help of some non-linear mapping (or kernel) functions, the input data is first transformed into a high-dimensional feature space, and classified according to the separating hyperplane, as shown in Figure 3.18.

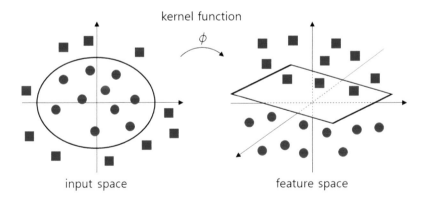

Figure 3.18: SVM concept. By mapping the original objects (left) to a higher dimensional feature space (right) using a kernel function, the two classes can now easily be separated. (Adapted from "Self-Diagnosis of Localization Status for Autonomous Mobile Robots", by Jiwoong Kim, Jooyoung Park, and Woojin Chung, 2018, Sensors 2018 18(9), p. 3168. ©Jiwoong Kim, Jooyoung Park, and Woojin Chung, https://www.mdpi.com/sensors/sensors-18-03168/article_deploy/html/images/sensors-18-03168-g006.png, "Kernel trick for mapping from an input space to a feature space", https://creative commons.org/licenses/by/4.0/legalcode)

3.4.2.2 Random forest

Random forest [7], as shown in Figure 3.19, is a collection of multiple decision trees that are automatically generated by random selection of data
and random selection of feature subsets. The classification result is determined by majority voting, i.e., taking the most popular result among the
results of all the decision trees. Compared to a single decision tree, random forest is more robust to overfitting (building a model that does not
generalize well) as it incorporates random noise as part of the model. It
also has lower variance due to the averaging effect of the trees [36].

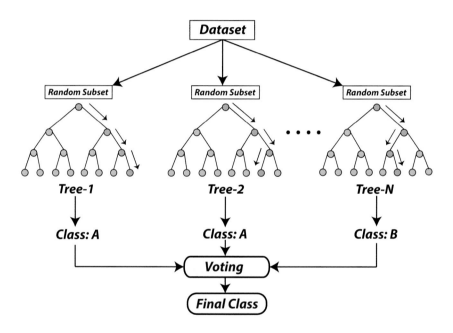

Figure 3.19: Random Forest algorithm. The input value is fed to a defined number of decision trees, each being generated using a random selection of the
dataset. The final (classification) output is selected by the majority vote of all
results.

3.4.2.3 Artificial neural network

An *artificial neural network (ANN)* is a multi-layered system of interconnected nodes (or *neurons*). Non-linear classification typically employs a special class of ANN, namely the *multi-layer perceptron (MLP)*. An MLP consists of at least three layers (input, hidden and output), where each feature is represented by a node in the input layer. An MLP is trained using the *backpropagation algorithm* [64], which repeatedly updates the weight of each node in the forward and backward direction until the correct classification is achieved, as shown in Figure 3.20. In the forward direction, the input data is multiplied with the actual weight of the node and, after applying some non-linear activation function, the output is propagated to the next layer until the final result from the output layer is obtained. The error between the actual and expected result is measured, and the weight of each node is adjusted accordingly starting from the output layer to the input layer (hence backpropagation) to reduce the error.

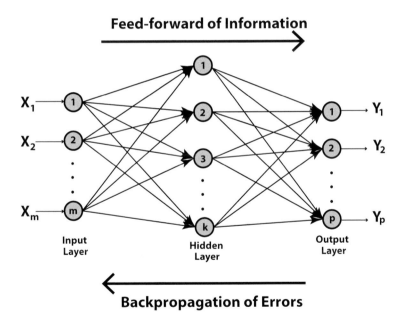

Figure 3.20: Multilayer perceptron with backpropagation concept.

An overview of classification algorithms can be found in [31]. Most of the popular classifiers (as well as other machine-learning algorithms) are publicly available as open-source libraries, such as mlpack [11] and OpenCV [47].

3.5   Multi-sensor data fusion

Multi-sensor data fusion was mentioned briefly earlier in this chapter. Multi-sensor data fusion is a method that combines the output of several sensors to create a more robust result.

GNSS technology, such as GPS, offers a simple and cost-effective way to get localized globally. However, it may not be accurate, and can even be unavailable in certain situations, such as indoor locations or in the urban canyon between skyscrapers. Therefore, a robust localization typically combines the result of multiple sensors using sensor data fusion. Using the available sensors and maps, the vehicle will be able to robustly perform the localization task.

Because there are a variety of sensors providing different sets of data, there needs to be a method to integrate, or fuse, the information. By combining data from these sensors, the uncertainty of the information from individual sensors can be reduced, as the weaknesses found in one sensor may be compensated by other sensor types. This, in turn, leads to a more reliable perception model and a better understanding of the external environment.

3.5.1   Classifications

The approaches to sensor fusion can be described using Durrant-Whyte's fusion classification based on the sensor relationship (see Figure 3.21) [18]:

- Complementary
  Complementary fusion means combining the partial information from two or more sensors to build a complete information of the observed object. There are several reasons why the sensor only gets partial information, including sensor defects, system limitations, partial object occlusion or simply due to the actual sensor placement in the vehicle.

- Redundant
  Sometimes sensors are intentionally configured such that their operating ranges overlap. Redundant fusion makes use of the

information from the overlapping areas to increase detection confidence or to act as a backup when one sensor fails.

■ Cooperative
Cooperative fusion takes information of different types to create new or more complex information. One example of cooperative fusion is combining angle and distance information to determine the location of an object [1].

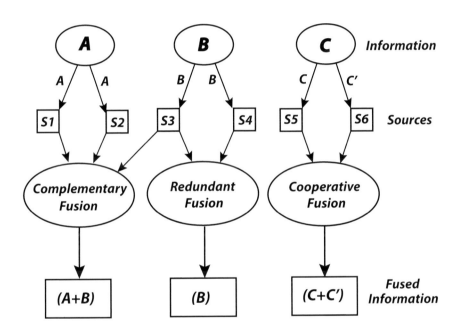

Figure 3.21: Durrant-Whyte's fusion classification. (Adapted from "A Review of Data Fusion Techniques" by Federico Castanedo, The Scientific World Journal, vol. 2013, Article ID 704504. ©2013 Federico Castenado)

Multi-sensor data fusion has its origin in military applications. Much of the pioneering work in this field was done by the Joint Directors of Laboratories (JDL) of the US Department of Defense in the late 1980s [65], which first introduced a multi-level data fusion model. This model is known as the JDL model and categorizes four fusion levels:

- Object assessment (level 1) refines the location, identity and other information of detected objects.

- Situation assessment (level 2) establishes the relationships between the objects and assigns the actual situation into pre-defined scenarios.

- Threat assessment (level 3) evaluates the impact of the detected scenario in level 2 and projects the possible outcome.

- Process assessment (level 4) improves the data fusion process of all levels.

Since its inception, there have been a lot of publications that propose revision, extension, or customization of the JDL model to make it more applicable to other domains. Ruser and Léon used the more generalized concept of *information fusion* and proposed three abstraction levels (see Figure 3.22) [50]:

- Signal level fuses the signals or raw data from multiple sensors. This level is also known as low-level data fusion in other literatures.

- Feature level fuses the extracted features from sensor data. This abstraction level is also sometimes referred as intermediate or characteristic-level fusion.

- Symbol level fuses the decisions or detection/classification results obtained from each sensor. Other popular terms for this level include high-level or decision-level fusion.

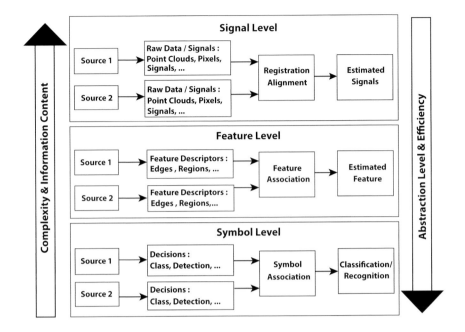

Figure 3.22: Sensor fusion abstraction levels. (Adapted from "Information fusion - An Overview" by Heinrich Ruser and Fernando Puente Léon, 2007, Tm-Technisches Messen, 74(3), p. 93-102)

Steinberg and Bowman [57] seek to make the model more general and applicable for civilian applications by using a less-militaristic term, such as *impact assessment* instead of *threat assessment*, and they added a new level *sub-object assessment* before the first fusion level (see Figure 3.23).

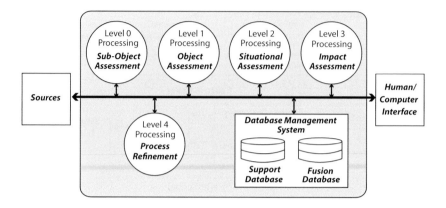

Figure 3.23: Revised Joint Directors of Laboratories (JDL) data fusion model. (Adapted from "Revisions to the JDL data fusion model", by Alan N. Steinberg, Alan N., Christopher L. Bowman, and Franklin E. White, 1999, Proc. SPIE Vol. 3719, Sensor Fusion: Architectures, Algorithms, and Applications III, p. 430-441)

Another well-known fusion classification system was introduced by Dasarathy [13]. The fusion classification is simply defined by the fusion's input and output types, which may not necessarily come from the same abstraction level. An overview of popular sensor fusion classifications can be found in [9].

3.5.2 Techniques

In the following sections we will discuss some of the popular techniques to solve the sensor data fusion problem.

3.5.2.1 Probabilistic approach

The classical approach for multi-sensor data fusion is to represent the uncertainty using probability models. The 'fused' data is then calculated using common methods for combining probability models, such as through direct Bayes inference or through recursive Bayesian filter. We will briefly examine the two methods in the following section.

Bayesian inference

Let us assume we want to fuse data from two observations z_1 and z_2, from the common state x. Using the probabilistic approach, we model the uncertainty of the two observations using probability density functions, $P(z_1)$ and $P(z_2)$, respectively. Even though the two observations are *obtained* independently, they are not totally independent as they share the common state x. Thus, the observations are said to be *conditionally independent* from the common state x, denoted as $P(z_1|x)$ and $P(z_2|x)$. Note that totally independent observations would mean that the observations are completely unrelated to each other and would make the fusion pointless.

Recall that Bayes' theorem is defined as follows:

$$P(A|B) = \frac{P(B|A)P(A)}{P(B)} \tag{3.17}$$

Applying Equation 3.17 to a set of M observations, $Z \triangleq z_1, z_2, \cdots, z_M$, the fused state x given all observations Z can be calculated as follows:

$$P(x|z_1, z_2, \cdots, z_M) = \frac{P(z_1, z_2, \cdots, z_m|x)P(x)}{P(z_1, z_2, \cdots, z_m)} \tag{3.18}$$

Hence the fusion result depends on the component $P(z_1, z_2, \cdots, z_m|x)$, the prior belief $P(x)$ and the normalizing factor $P(z_1, z_2, \cdots, z_m)$. Whereas the latter can be calculated easily, the two former terms, i.e., the joint probability distribution of all observations and the prior, are unknown. There are several common approaches to address this issue, usually by making some assumption about the underlying probability distributions.

The *independent opinion pool* approach is based upon the assumption that, apart from being conditionally independent to the common state x, the observations are independent to each other [4]. The independent opinion pool is stated in Equation 3.19 as follows:

$$P(x|z_1, z_2, \cdots, z_M) \propto \prod_{m=1}^{M} P(x|z_m) \tag{3.19}$$

If the prior information seems to be from the same origin, using the following *independent likelihood pool*, as described in Equation 3.20, is more appropriate [4]:

$$P(x|z_1, z_2, \cdots, z_M) \propto P(Z) \prod_{m=1}^{M} P(z_m|x) \tag{3.20}$$

Last but not least, the *linear opinion pool* (see Equation 3.21) should be preferred if there are dependencies between the sensors. Unlike the previous opinion pools, the linear opinion approach is stated as the sum of weighted individual posteriors. This way the reliability of each sensor can be expressed as its weight w_m. Noisy sensors are assigned with lower weights, so that more reliable sensors play a larger role in defining the final result.

$$P(x|z_1, z_2, \cdots, z_M) = \sum_{m=1}^{M} w_m P(x|z_m) \qquad (3.21)$$

Recursive Bayesian filtering

One major disadvantage of the Bayesian inference method is that all past observations in the history are needed for the posterior calculation. An alternative approach is to rearrange the last equation into a recursive filter as Equation 3.22 follows:

$$P(x|Z_t) = P(x|Z_t)P(Z_t) = P(z_t|x)P(Z_{t-1})P(x) \qquad (3.22)$$

Because $Z_t \triangleq z_t, Z_{t-1}$ and so by applying Bayes' chain rule $P(z_t|Z_{t-1}) = \frac{P(Z_t)}{P(Z_{t-1})}$, we get the following Equation 3.23:

$$P(x|Z_t) = \frac{P(z_t|x)P(x|Z_{t-1})}{P(z_t|Z_{t-1})} \qquad (3.23)$$

Note the recursiveness of the Equation 3.23. The posterior becomes the new prior for the next iteration. This way we do not have to deal with all observations in the history, since they are now *summarized* in the term $P(x|Z_{t-1})$. Thus, in contrast to the Bayesian inference method, the complexity does not grow with the number of observations.

3.5.2.2 *Evidential approach*

The evidential approach is based on the *Dempster-Shafer (DS)* theory of evidence, first introduced by Arthur P. Dempster in the 1960s [14] and further developed about a decade later by Glenn Shafer [53]. The major limitation of probabilistic approaches is that they can only model one type of uncertainty: the so-called *aleatory uncertainty*, sometimes simply known as variability or randomness. In order to estimate the likelihood of a hypothesis using the probabilistic approach, we need to know, or assume, the prior and conditional probabilities of the probabilistic model. The knowledge is usually derived from statistical data, laws of physics, or, sometimes, is just assumed using common sense. However, such knowledge might be inaccurate, too complicated to model, or incomplete, i.e., there are some random variables that affect the overall likelihood of a hypothesis but cannot be accounted for in the model because they're unknown. The evidential approach is based on a different method of reasoning under uncertainty, namely, computing the likelihood of a hypothesis based on the probability of the evidence that supports it. Hence, the *epistemic uncertainty*, also known as the systematic uncertainty or the uncertainty due to incomplete/lack of knowledge of the full system, can also be taken into account. Another advantage of the evidential approach is the ability to model ignorance, so that *lack of belief* is easily distinguishable from *disbelief* [35].

The DS theory uses a set of *mass functions* to represent the degree of belief that supports each proposition. The mass function, also known as the *Basic Belief Assignment (BBA)*, is a function $m(\cdot): 2^\Omega \rightarrow [1,0]$ that satisfies the following constraints:

$$m(\emptyset) = 0 \tag{3.24}$$

and

$$\sum_{A \in 2^\Omega} m(A) = 1 \tag{3.25}$$

where Ω is the set of mutually exhaustive and exclusive hypotheses, also known as the *Frame of Discernment (FoD)*. For example, in the case of an occupancy grid map, where each cell is either free or occupied $\Omega = \{Free, Occupied\}$. The *power set* is defined as $2^\Omega = 4$, such that each cell is represented by the following mass functions: $m(\emptyset)$, $m(Free)$, $m(Occupied)$, $m(Free, Occupied)$. However, due to the first constraint $m(\emptyset) = 0$, there are effectively only three mass functions to consider. The last mass function represents the epistemic uncertainty, sometimes denoted by $m(Conflict)$ or $m(dontknow)$ in some literatures.

In DS theory, a *confidence interval* is bounded by *belief* (or credibility) and *plausibility* functions defined in Equations 3.26 and 3.27 as follows:

$$Bel(A) = \sum_{\substack{B \subseteq A \\ B \in G^\Omega}} m(B) \tag{3.26}$$

$$Pl(A) = \sum_{\substack{B \cap A \neq \emptyset \\ B \in G^\Omega}} m(B) \tag{3.27}$$

The confidence interval of proposition A is given as $[\text{Bel}(A), \text{Pl}(A)]$, where $0 \leqslant \text{Bel}(A) \leqslant \text{Pl}(A) \leqslant 1$. A confidence interval of $[1,1]$ implies that the proposition is true based on all the evidence. In contrast, a confidence interval of $[0,0]$ means that the hypothesis is false based on all available evidence. A confidence interval of $[0,1]$ denotes a complete ignorance of the proposition as no evidence supports or disproves it.

In a nutshell, sensor data fusion based on the evidential approach involves building the mass functions over all propositions in the fusion space, i.e., all propositions in the power-set 2^Ω, updating them incrementally based on new sensor observations and computing the joint mass function by applying some fusion rule. The standard fusion rule for combining multiple sources of information is *Dempster's rule of combination* or the *DS rule* [53], as defined in the following Equation 3.28:

$$m_{1,2}^{DS}(X) = [m_1 \oplus m_2](X) = \frac{m_{1,2}(X)}{1 - K_{1,2}} \tag{3.28}$$

where:

$$m_{1,2}(X) \triangleq \sum_{\substack{X_1, X_2 \in 2^\Omega \\ X_1 \cap X_2 = X}} m_1(X_1)\, m_2(X_2) \tag{3.29}$$

where $K_{1,2}$ is called the *degree of conflict* as defined in Equation 3.30 belows:

$$K_{1,2}(X) \triangleq m_{1,2}(\emptyset) = \sum_{\substack{X_1, X_2 \in 2^\Omega \\ X_1 \cap X_2 = \emptyset}} m_1(X_1)\, m_2(X_2) \tag{3.30}$$

Note that the DS rule is mathematically undefined if the two sources are in total conflict, i.e., $K_{1,2} = 1$. The DS rule is also known to cope badly with both high and low conflicting situations [15]. Therefore, some alternative rules have been proposed to overcome these limitations. One of the popular alternatives is the *Probabilistic Conflict Redistribution Rule No. 6 (PCR6)*, whose basic idea is to redistribute conflicting masses to the non-empty elements involved in the conflict [41].

Having calculated the fused mass functions, a decision can now be made based on some criteria. There are several criteria for choosing the best hypothesis for the fusion result. For example, we can take the proposition with the highest degree of belief/credibility, or the one with the highest plausibility, or the one with the highest likelihood after some probabilistic transformation, i.e., being translated from the belief function model to the probability model. A popular transformation method is the *pignistic probabilistic transformation* proposed by Philippe Smets in the early 1990s and defined as follows [54]:

$$P\{A\} = \sum_{X \in 2^\Omega} \frac{X \cap A}{X} \frac{m(X)}{1 - m(\emptyset)} = \sum_{X \in 2^\Omega} \frac{X \cap A}{X} m(X) \tag{3.31}$$

where $m(\emptyset) = 0$ according to one of the two DS constraints mentioned in the beginning of this section.

3.5.2.3 *Other approaches*

Besides the probabilistic and evidential approaches, other approaches have been proposed to solve the sensor data fusion problem. However, they are less popular in the context of SDV. One such alternative approach is based on the *fuzzy set theory*, also known as the *possibilistic* approach [48].

3.6 Summary

As we have seen in this chapter, perception is the single most important function of an SDV. Without the ability to monitor the surroundings and work out where it is, an SDV could never know how to reach its destination safely, avoiding hazards and finding a suitable route.

As we saw, the perception task consists of many subtasks, including localization, mapping, object detection, and multi-sensor data fusion. Often for efficiency, the first two depend closely on each other and are usually solved simultaneously as SLAM or simultaneous location and mapping. Mapping is about knowing the terrain, roads, etc. that surround you and localization is about knowing exactly where the vehicle sits in relation to the map. In other words, the functions can be summed up as, 'where am I, and what direction am I facing?'

We introduced the two main SLAM approaches in Section 3.3. The filtering approach uses either Gaussian Bayesian filters such as the Kalman filter or non-Gaussian filters such as the particle filter. In the filtering approach, the current position is estimated by combining the most recent observations with past predictions of your position, taking account of the likely source of error (noise). The optimization approach takes a slightly different approach. Here, all observations from the start are combined and smoothed to find the most likely overall trajectory that is consistent with the set of observations. In graph-based SLAM, a pose-constraint graph is constructed and the task is to find the configuration that is maximally consistent with that graph. In bundle adjustment, the aim is to find the optimal configuration that minimizes the cost function between observed features and the expected projection of those features.

By contrast, object detection is about identifying objects in your vicinity that may pose a hazard, in particular moving objects (DATMO or detection of moving objects). This is about answering the question, 'How can I avoid collision with another object?'. As explained in Section 3.4, object detection consists of three main sub-problems: object localization, object classification and semantic segmentation. The common approach to achieve this requires a combination of feature extraction and classification. Feature detection can be done using techniques such as histogram of oriented gradients, scale-invariant feature transform or maximally stable extremal regions. All these approaches try to extract the intrinsic characteristics that allow a given feature to be distinguishable from other features. Classification then seeks to compare these features to known features in order to classify them into one of several categories, e.g., pedestrian, vehicle, footpath, etc. The main techniques here are support vector machine, random forest and artificial neural networks.

Key to all these tasks is the interpretation of the data from the multiple sensors on the vehicle. As we saw, the raw data from these sensors has to be interpreted and combined with other data to be any use. Multi-sensor data fusion is the term for a series of techniques that are designed to provide more robust sensing by combining the outputs from several sensors in a complementary, redundant or cooperative fashion. Data fusion techniques are either probabilistic, where techniques such as Bayesian inference are used to predict the most likely combined result based on the uncertainty in the data; or evidential, where the aim is to find which of a set of possible combined results is the most likely given the observations.

In the next chapter, we will see how the SDV's software stack allows it to take this knowledge of the environment and use it to drive the vehicle to its destination. We will look at how all the software fits into a combined architecture and will explore a couple of well-known middlewares used for SDVs.

References

[1] Ahmed Abdelgawad and Magdy Bayoumi. Data fusion in wsn. In *Resource-Aware Data Fusion Algorithms for Wireless Sensor Networks*, pages 17–35. Springer, 2012.

[2] Sameer Agarwal, Noah Snavely, Steven M Seitz, and Richard Szeliski. Bundle adjustment in the large. In *European Conference on Computer Vision*, pages 29–42. Springer, 2010.

[3] Thomas Bengtsson, Peter Bickel, Bo Li, et al. Curse-of-dimensionality revisited: Collapse of the particle filter in very large scale systems. In *Probability and Statistics: Essays in Honor of David A. Freedman*, pages 316–334. Institute of Mathematical Statistics, 2008.

[4] James O Berger. *Statistical Decision Theory and Bayesian Analysis*. Springer Science & Business Media, 2013.

[5] Paul J Besl and Neil D McKay. Method for registration of 3-d shapes. In *Sensor Fusion IV: Control Paradigms and Data Structures*, volume 1611, pages 586–607. International Society for Optics and Photonics, 1992.

[6] Bernhard E Boser, Isabelle M Guyon, and Vladimir N Vapnik. A training algorithm for optimal margin classifiers. In *Proceedings of the Fifth Annual Workshop on Computational Learning Theory*, pages 144–152. ACM, 1992.

[7] Leo Breiman. Random forests. *Machine Learning*, 45(1):5–32, 2001.

[8] Ramon F Brena, Juan Pablo García-Vázquez, Carlos E Galván-Tejada, David Muñoz-Rodriguez, Cesar Vargas-Rosales, and James

Fangmeyer. Evolution of indoor positioning technologies: A survey. *Journal of Sensors*, 2017. https://www.hindawi.com/journals/js/2017/2630413/.

[9] Federico Castanedo. A review of data fusion techniques. *The Scientific World Journal*, 2013. https://www.hindawi.com/journals/tswj/2013/704504/.

[10] Andrea Censi. An icp variant using a point-to-line metric. In *Robotics and Automation, 2008. ICRA 2008. IEEE International Conference on*, pages 19–25. IEEE, 2008.

[11] Ryan R Curtin, James R Cline, Neil P Slagle, William B March, Parik-shit Ram, Nishant A Mehta, and Alexander G Gray. Mlpack: A scalable c++ machine learning library. *Journal of Machine Learning Research*, 14(Mar):801–805, 2013.

[12] Navneet Dalal and Bill Triggs. Histograms of oriented gradients for human detection. In *Computer Vision and Pattern Recognition, 2005. CVPR 2005. IEEE Computer Society Conference on*, volume 1, pages 886–893. IEEE, 2005.

[13] Belur V Dasarathy. Sensor fusion potential exploitation-innovative architectures and illustrative applications. *Proceedings of the IEEE*, 85(1):24–38, 1997.

[14] Arthur P Dempster. Upper and lower probabilities induced by a multivalued mapping. In *Classic Works of the Dempster-Shafer Theory of Belief Functions*, pages 57–72. Springer, 2008.

[15] Jean Dezert, Pei Wang, and Albena Tchamova. On the validity of dempster-shafer theory. In *Information Fusion (FUSION), 2012 15th International Conference on*, pages 655–660. IEEE, 2012.

[16] Randal Douc and Olivier Cappé. Comparison of resampling schemes for particle filtering. In *Image and Signal Processing and Analysis, 2005. ISPA 2005. Proceedings of the 4th International Symposium on*, pages 64–69. IEEE, 2005.

[17] Arnaud Doucet, Nando De Freitas, Kevin Murphy, and Stuart Russell. Rao-blackwellised particle filtering for dynamic bayesian networks. In *Proceedings of the Sixteenth Conference on Uncertainty in Artificial Intelligence*, pages 176–183. Morgan Kaufmann Publishers Inc., 2000.

[18] Hugh F Durrant-Whyte. Sensor models and multisensor integration. In *Autonomous Robot Vehicles*, pages 73–89. Springer, 1990.

[19] Jakob Engel, Thomas Schöps, and Daniel Cremers. Lsd-slam: Large-scale direct monocular slam. In *European Conference on Computer Vision*, pages 834–849. Springer, 2014.

[20] Jakob Engel, Jörg Stückler, and Daniel Cremers. Large-scale direct slam with stereo cameras. In *Intelligent Robots and Systems (IROS), 2015 IEEE/RSJ International Conference on*, pages 1935–1942. IEEE, 2015.

[21] Li Fei-Fei. Object recognition. https://vision.stanford.edu/documents/Fei-Fei_ICVSS07_ObjectRecognition_web.pdf. [accessed 03-Oct-2018].

[22] Juan-Antonio Fernández-Madrigal. *Simultaneous Localization and Mapping for Mobile Robots: Introduction and Methods: Introduction and Methods*. IGI Global, 2012.

[23] Friedrich Fraundorfer and Davide Scaramuzza. Visual odometry: Part ii: Matching, robustness, optimization, and applications. *IEEE Robotics & Automation Magazine*, 19(2):78–90, 2012.

[24] Dorian Gálvez-López and Juan D Tardos. Bags of binary words for fast place recognition in image sequences. *IEEE Transactions on Robotics*, 28(5):1188–1197, 2012.

[25] Giorgio Grisetti. Notes on least-squares and slam draft. 2014. http://www.dis.uniroma1.it/~grisetti/teaching/lectures-ls-slam-master_2015_16/web/reading_material/grisetti12stest.pdf [accessed 03-Oct-2018].

[26] Giorgio Grisetti, Rainer Kummerle, Cyrill Stachniss, and Wolfram Burgard. A tutorial on graph-based slam. *IEEE Intelligent Transportation Systems Magazine*, 2(4):31–43, 2010.

[27] Fredrik Gustafsson. Particle filter theory and practice with positioning applications. *IEEE Aerospace and Electronic Systems Magazine*, 25(7):53–82, 2010.

[28] Dirk Hahnel, Wolfram Burgard, Dieter Fox, and Sebastian Thrun. An efficient fastslam algorithm for generating maps of large-scale cyclic environments from raw laser range measurements. In *Intelligent Robots and Systems, 2003.(IROS 2003). Proceedings. 2003 IEEE/RSJ International Conference on*, volume 1, pages 206–211. IEEE, 2003.

[29] John H Halton. Sequential monte carlo techniques for solving non-linear systems. *Monte Carlo Methods and Applications MCMA*, 12(2):113–141, 2006.

[30] Kurt Konolige and Willow Garage. Sparse sparse bundle adjustment. In *BMVC*, volume 10, pages 102–1. Citeseer, 2010.

[31] Sotiris B Kotsiantis, I Zaharakis, and P Pintelas. Supervised machine learning: A review of classification techniques. *Emerging artificial intelligence applications in computer engineering*, 160:3–24, 2007.

[32] Rainer Kümmerle, Giorgio Grisetti, Hauke Strasdat, Kurt Konolige, and Wolfram Burgard. g 2 o: A general framework for graph optimization. In *Robotics and Automation (ICRA), 2011 IEEE International Conference on*, pages 3607–3613. IEEE, 2011.

[33] JM Lee. Introduction to Smooth Manifolds, Graduate Texts in Mathematics, series volume 218, Springer, 2003.

[34] Jun S Liu. Metropolized independent sampling with comparisons to rejection sampling and importance sampling. *Statistics and Computing*, 6(2):113–119, 1996.

[35] Liping Liu. A theory of gaussian belief functions. *International Journal of Approximate Reasoning*, 14(2-3):95–126, 1996.

[36] Gilles Louppe. Understanding random forests: From theory to practice. *arXiv preprint arXiv:1407.7502*, 2014.

[37] Manolis Lourakis. Bundle adjustment gone public. *PRCV Colloquium Prague*, 2011. http://users.ics.forth.gr/~lourakis/sba/PRCV_colloq.pdf

[38] MLA Lourakis and Antonis A Argyros. Is Levenberg-Marquardt the most efficient optimization algorithm for implementing bundle adjustment? In *Computer Vision, 2005. ICCV 2005. Tenth IEEE International Conference on*, volume 2, pages 1526–1531. IEEE, 2005.

[39] David G Lowe. Distinctive image features from scale-invariant keypoints. *International Journal of Computer Vision*, 60(2):91–110, 2004.

[40] Feng Lu and Evangelos Milios. Robot pose estimation in unknown environments by matching 2d range scans. *Journal of Intelligent and Robotic Systems*, 18(3):249–275, 1997.

[41] Arnaud Martin and Christophe Osswald. A new generalization of the proportional conflict redistribution rule stable in terms of decision. *Advances and Applications of DSmT for Information Fusion: Collected Works Volume 2*, 2:69–88, 2006.

[42] Jiri Matas, Ondrej Chum, Martin Urban, and Tomás Pajdla. Robust wide-baseline stereo from maximally stable extremal regions. *Image and Vision Computing*, 22(10):761–767, 2004.

[43] Michael Montemerlo, Sebastian Thrun, Daphne Koller, Ben Wegbreit, et al. Fastslam: A factored solution to the simultaneous localization and mapping problem. *Aaai/iaai*, 593598, 2002.

[44] Etienne Mouragnon, Maxime Lhuillier, Michel Dhome, Fabien Dekeyser, and Patrick Sayd. Real time localization and 3d reconstruction. In *Computer Vision and Pattern Recognition, 2006 IEEE Computer Society Conference on*, volume 1, pages 363–370. IEEE, 2006.

[45] Raul Mur-Artal and Juan D Tardós. Orb-slam2: An open-source slam system for monocular, stereo, and rgb-d cameras. *IEEE Transactions on Robotics*, 33(5):1255–1262, 2017.

[46] Edwin B Olson. Real-time correlative scan matching. *Ann Arbor*, 1001:48109, 2009.

[47] OpenCV. Open source computer vision library. https://opencv.org/. [accessed 03-Oct-2018].

[48] Giuseppe Oriolo, Giovanni Ulivi, and Marilena Vendittelli. Real-time map building and navigation for autonomous robots in unknown environments. *IEEE Transactions on Systems, Man, and Cybernetics, Part B (Cybernetics)*, 28(3):316–333, 1998.

[49] Ethan Rublee, Vincent Rabaud, Kurt Konolige, and Gary Bradski. Orb: An efficient alternative to sift or surf. In *Computer Vision (ICCV), 2011 IEEE International Conference on*, pages 2564–2571. IEEE, 2011.

[50] Heinrich Ruser and Fernando Puente Léon. Informationsfusion-eine übersicht (information fusion-an overview). *tm-Technisches Messen*, 74(3):93–102, 2007.

[51] Ehab Salahat and Murad Qasaimeh. Recent advances in features extraction and description algorithms: A comprehensive survey. In *Industrial Technology (ICIT), 2017 IEEE International Conference on*, pages 1059–1063. IEEE, 2017.

[52] Matthias Schreier. Environment representations for automated on-road vehicles. *at-Automatisierungstechnik*, 66(2):107–118, 2018.

[53] Glenn Shafer. *A mathematical Theory of Evidence*, volume 42. Princeton University Press, 1976.

[54] Philippe Smets. Constructing the pignistic probability function in a context of uncertainty. In *UAI*, volume 89, pages 29–40, 1989.

[55] Randall Smith, Matthew Self, and Peter Cheeseman. Estimating uncertain spatial relationships in robotics. In *Autonomous robot vehicles*, pages 167–193. Springer, 1990.

[56] Cyrill Stachniss, John J. Leonard, and Sebastian Thrun. *Simultaneous Localization and Mapping*, pages 1153–1176. Springer International Publishing, Cham, 2016.

[57] Alan N Steinberg and Christopher L Bowman. Revisions to the jdl data fusion model. In *Handbook of Multisensor Data Fusion*, pages 65–88. CRC Press, 2008.

[58] Sebastian Thrun. Probabilistic robotics. *Communications of the ACM*, 45(3):52–57, 2002.

[59] Gian Diego Tipaldi and Kai O Arras. Flirt-interest regions for 2d range data. In *Robotics and Automation (ICRA), 2010 IEEE International Conference on*, pages 3616–3622. IEEE, 2010.

[60] Bill Triggs, Philip F. McLauchlan, Richard I. Hartley, and Andrew W. Fitzgibbon. Bundle adjustment - A modern synthesis. In *Proceedings of the International Workshop on Vision Algorithms: Theory and Practice*, ICCV '99, pages 298–372, London, UK, 2000. Springer-Verlag.

[61] Andrea Vedaldi and Brian Fulkerson. Vlfeat: An open and portable library of computer vision algorithms. In *Proceedings of the 18th ACM International Conference on Multimedia*, pages 1469–1472. ACM, 2010.

[62] Damien Vivet, Paul Checchin, and Roland Chapuis. Localization and mapping using only a rotating fmcw radar sensor. *Sensors*, 13(4):4527–4552, 2013.

[63] Erik Ward and John Folkesson. Vehicle localization with low cost radar sensors. In *Intelligent Vehicles Symposium (IV), 2016 IEEE*. Institute of Electrical and Electronics Engineers (IEEE), 2016.

[64] Paul Werbos. Beyond regression: new tools for prediction and analysis in the behavioral sciences. *Ph. D. dissertation, Harvard University*, 1974.

[65] Franklin E White et al. A model for data fusion. In *Proc. 1st National Symposium on Sensor Fusion*, volume 2, pages 149–158, 1988.

[66] H Durrant Whyte. Simultaneous localisation and mapping (slam): Part i the essential algorithms. *Robotics and Automation Magazine*, 2006.

[67] Nan Yang, Rui Wang, and Daniel Cremers. Feature-based or direct: An evaluation of monocular visual odometry. *arXiv preprint arXiv:1705.04300*, 2017.

[68] Faheem Zafari, Athanasios Gkelias, and Kin Leung. A survey of indoor localization systems and technologies. *arXiv preprint arXiv:1709.01015*, 2017.

Chapter 4

Architecture

In the previous chapter, we discussed how SDVs perceive the environment around them through a combination of localization, mapping and object detection. In this chapter we explore how the SDV combines this knowledge of its environment with other data such as its destination, the rules of the road, and its knowledge of its own capabilities in order to achieve the aim of driving itself safely to the destination.

SDV software can be viewed from two perspectives. The functional architecture view considers the actual functions that the software needs to perform. These include perception, localization, mapping and object detection which we explored in Chapter 3. In this chapter we will look at the other functions an SDV needs in order to work. These include planning and vehicle control. The system architecture view looks at how these discrete functions can be combined in order to create a system that is capable of the required level of autonomous driving. We start with exploring the functional architecture before moving on to looking at the system architecture. At the end of the chapter we will describe some real-world examples of SDV middleware.

4.1   Functional architecture

The functional architecture consists of three main parts: perception (already discussed), planning and vehicle control. In this section we will explore how these three main functions allow an SDV to know where it is and to drive itself safely to its destination.

4.1.1 *Perception*

Perception is all about answering the questions, 'where am I?' and 'what is happening around me?'. The main perception functions are localization, mapping and object detection. These functions were described in detail in the previous chapter, but we will summarize them very briefly here.

As we saw in Section 3.1, localization allows a vehicle to position itself within the map of its environment. This includes knowing its orientation within the map. Localization relies heavily on sensors such as GNSS, IMUs, lidar, cameras or odometry. Typical approaches to localization rely on either scan matching, where you match your view of the environment with the map, or on dead-reckoning, where you use your knowledge of heading, speed and time to track from a known location on the map to a new location.

Mapping is the process used to construct an accurate picture of your surroundings (see Section 3.2). This may be based on a global map or it may be purely local. The important thing is that the map must be extremely accurate for an SDV to be able to operate safely. There are three popular classes of map: Occupancy Grid Maps, Feature Maps and Relational Maps. All three types have pros and cons, and the particular choice may be dictated by the environment and the constraints of the SDV.

Mapping and localization are heavily co-dependent, and often a vehicle has neither an accurate map nor an accurate location. This is where simultaneous localization and mapping (see Section 3.3) is used. SLAM algorithms seek to build a map of the environment at the same time as trying to position the vehicle within that map. If the algorithm recognizes that it has returned to a location it already visited, then it can perform loop closure. The main forms of SLAM are the filtering approach, using either Kalman or particle filters, and the optimization approach, using either graph-based techniques or bundle adjustment.

The final perception function is object detection (see Section 3.4). This is critical for any SDV to be able to safely navigate because it allows it to detect moving hazards such as pedestrians and other vehicles. Object detection consists of processing an image, extracting features within that image and classifying those features into objects in order to create a semantic representation (or map). The main feature extraction techniques we looked at were HOG, SIFT and MSER. Classification can be performed using Support Vector Machine, Random Forest, Artificial Neural Network or other machine-learning algorithms.

4.1.2 Planning

Planning involves answering the question, 'how do I get to my destination?'.

The planning activities can be described by using a top-down approach containing a hierarchy of three layers: route, behavioral, and motion planning, all described below.

4.1.2.1 Route planning

At this level, the SDV performs the calculation to determine the best route to travel from the current location to the destination based on the road network information provided by a map. The route calculation might need to consider other external factors, such as real-time traffic information, the estimated energy consumption (especially for electric vehicles), the user's preference whether or not to use toll roads, etc. This level of planning is also performed by built-in car navigation systems, after-market navigation systems (TomTom, Navigon, etc.), and mobile apps (Google Maps, Here, etc.).

Route planning typically uses special algorithms for solving a common graph theory problem called the *shortest path problem*. The problem can be defined as finding the shortest path between two nodes in a graph. One of the best-known shortest path algorithms is *Dijkstra's algorithm* [6] as shown in Figure 4.1. The algorithm starts by initializing the distance value of all nodes to infinity. For all directly reachable nodes from the starting node a new distance value or cost is calculated, and the value is updated if the distance is shorter. This process iterates through the entire graph until all nodes have been traversed. The shortest path to any destination can now be determined by summing the cost of the node and the set of registered edges to reach that node. Faster algorithms, such as Contraction Hierarchies [9], perform some precomputation steps to speed up the process.

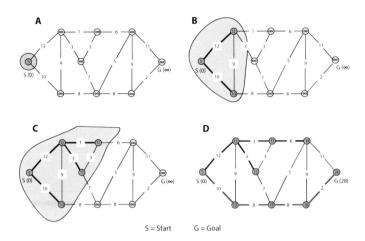

Figure 4.1: Dijkstra's algorithm. (A) Initialize all costs to infinity. (B) The cost to the direct neighbors is updated. (C) Second iteration. The costs are updated. (D) The shortest path from the start node to any given node in the graph can be determined. In this instance, the shortest path between the two outermost nodes is 28 and the resulting path is highlighted in red.

Several implementations of state-of-the-art route planning algorithms are freely available as open source projects, e.g., RoutingKit [4] or GrassHopper [10].

4.1.2.2   *Behavioral planning*

The result of the previous route planning step is a set of road network segments or waypoints that the SDV needs to follow from its current position. The next planning step is behavioral planning, which decides how best to reach the next waypoint under the actual local driving context, i.e., with regard to the current road geometry, perceived obstacles, other traffic participants, actual traffic rules (speed limit, no passing zone), limitation of vehicle control, etc. The result of this planning step is a high-level decision, such as changing lane, lane following, merging, overtaking, etc.

One of the most challenging issues in behavior planning involves predicting the behavior of dynamic objects in the environment. This is especially critical for mixed traffic environments where SDVs share the road with normal vehicles. There are several proposed approaches to address the issue of decision-making under the uncertainty of other traffic participants' behaviors. The Prediction and Cost-function Based (PCB) approach [17] (see Figure 4.2) generates multiple candidates for possible longitudinal and lateral control directives, uses a prediction engine to forward-simulate the directives to generate the simulated trajectory and predicts the reaction of the surrounding vehicles in each simulation step and then selects the best decision by evaluating the total cost of progress, comfort, safety and fuel consumption. Claussmann et al. [2] provide an overview and comparison of state-of-the-art AI-based approaches, categorized into knowledge-based inference engine approaches, heuristic algorithms, approximate reasoning and human-like decision-making methods, as shown in Figure 4.3.

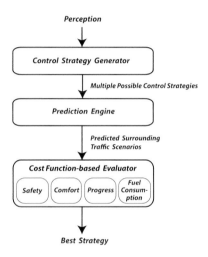

Figure 4.2: Block diagram of the PCB algorithm. (Adapted from "A behavioral planning framework for autonomous driving", by Junqing Wei, Jarrod M. Snider, Tianyu Gu, John M. Dolan, Bakhtiar Litkouhi, 2014, 2014 IEEE Intelligent Vehicles Symposium Proceedings, p. 458-464)

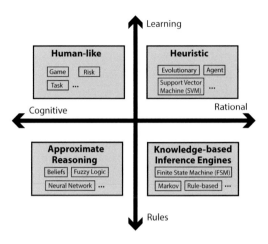

Figure 4.3: Classification of AI-based state-of-the-art behavioral planning approaches. (Adapted from "A study on al-based approaches for high-level decision making in highway autonomous driving", by Laurène Claussmann, Marc Revilloud, Sébastien Glaser, Dominique Gruyer, 2017, 2017 IEEE International Conference on Systems, Man, and Cybernetics, p. 3671-3676 ©2017 IEEE)

4.1.2.3   Motion planning

The high-level decision from the previous behavior planning step is executed in the motion planning step. The result of this last planning step is a set of driving commands for the vehicle controller over time, typically in the form of steering wheel angle, brake and throttle level. In robotics literature, the problem of motion planning is generally divided into two sub-problems: *path planning* and *trajectory planning*. Path planning is the task of finding the shortest collision-free geometric path from the start point to the destination. Trajectory planning is the task of determining the sequence of motions, as a function of time, to achieve a smooth drive along the desired path. Hence a path can also be regarded as a set of trajectories with a specification of the vehicle's velocity, acceleration and sometimes jerk (change of acceleration) at each point.

In order to find the best geometric path in path planning, the vehicle's environment generated by the map information needs to be combined with the information perceived by the sensors and other sources in the form of a discrete representation. Suitable representations include occupancy grids or driving corridors. As explained in the previous chapter, in occupancy grids, the vehicle's environment is divided into 2D grid cells. Each cell in the grid contains the probability of that cell being occupied by an obstacle. Driving corridors represent the free space in which the vehicle can travel without collision by taking into account all detected obstacles as well as other physical boundaries, such as the allowed lane and road boundaries. Each representation has its advantages and disadvantages. While the construction of occupancy grids is generally not computationally intensive, they tend to require more memory as the memory consumption increases proportionally to the total number of cells in the grid and the resolution of the grid. On the other hand, driving corridors always provide continuous collision-free space for the vehicle to move, but are generally more computationally intensive to construct. The advantages and disadvantages of different discrete representations are described in [12].

Motion planning is one of the most well-researched fields in robotics and there have been many algorithms developed since the late 1960s to solve challenging planning problems. However, not all of these algorithms are directly applicable for SDVs. The driving environment of an SDV is highly dynamic due to moving obstacles and also involves the complexity of vehicle dynamics at high speed.

Two of the popular path planning algorithms used for SDVs are the Probabilistic Roadmap (PRM) [13] and the Rapidly-exploring Random Tree (RRT) [15]. Both algorithms are examples of the so-called *sampling-based planning* approach. The sampling-based approach uses random node probing for the path exploration and a fast collision avoidance function to validate the new candidate nodes. As opposed to the *combinatorial planning* approach, the sampling-based approach does not require the full construction of all possible collision-free paths between the start and the goal node. Thus, the sampling-based approach might give a suboptimal solution, but it is more practical, especially when the search space is highly dimensional. The PRM algorithm, as illustrated in Figure 4.4, starts with knowledge of the start and end points, and the constraints. The algorithm randomly distributes a defined number of nodes in the space and connects each node to its nearest neighbor. All nodes and edges that connect the nodes must be collision-free according to a collision avoidance function. After all the nodes are connected, the resulting path between the start and the goal nodes can be easily determined using a shortest path algorithm such as Dijkstra. As shown in Figure 4.5, the RRT algorithm works by choosing a random node and trying to connect it to the current tree. The collision avoidance function validates whether the new node is collision-free. It continues until the goal node is reached. The resulting path is the set of edges that connect the start and the goal nodes in the tree.

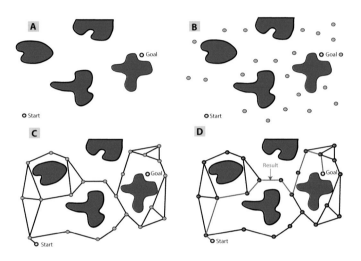

Figure 4.4: PRM algorithm. (A) Example of a configuration space showing the start and goal positions. (B) The algorithm starts by randomly distributing a defined number of collision-free nodes throughout the configuration space. (C) All nodes are connected by collision-free edges to their nearest neighbors. (D) The resulting path can be determined, e.g., by applying Dijkstra's shortest path algorithm to the graph.

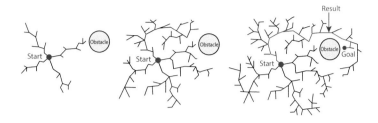

Figure 4.5: RRT algorithm. (Left) Some new nodes are generated randomly and connected to the start node. (Center) Some other nodes are created randomly and all collision-free ones are connected to the closest available node. (Right) The algorithm ends when the start and goal nodes are interconnected or the maximum iteration limit has been reached. (Adapted from "Rapidly-Exploring Random Trees: A New Tool for Path Planning", by Steven M. LaValle, 1998)

Trajectory planning mainly deals with finding the optimal geometric curve such as a spline, Bezier curve, or clothoid, that ensures a smooth motion through the desired path under the consideration of the current vehicle dynamics as illustrated in Figure 4.6. The result is the optimized mathematical expression to define that geometric curve. There have been several approaches proposed for this, the most popular ones include the Genetic Algorithm and the Sequential Quadratic Programming (SQP). In [16] these two approaches are explained, and their performances are compared.

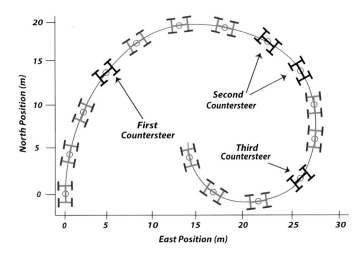

Figure 4.6: Example of a vehicle trajectory plan.

The Open Motion Planning Library (OMPL) [14] is an open source library that contains implementations of RRT, PRM and many other state-of-the-art sampling-based algorithms. As it is designed as a pure collection of planning algorithms, any additional functionalities, such as the collision avoidance function or visualization, still need to be implemented or provided by the SDV middleware.

4.1.3   Vehicle control

The main goal of vehicle control is to execute the decision made in the previous steps while ensuring the safe operation of the vehicle. Vehicle control typically involves translating the calculated trajectory into a set of control commands for the actuators, ensuring the stability of the vehicle and minimizing the impact of unexpected events. The latter is crucial, given that the probability of sensor/hardware failures, measurement incorrectness, or even implementation error, is never zero.

Due to their higher safety requirements (typically ASIL C or D), vehicle control modules are normally implemented and handled separately from the other SDV applications. The independence from other SDV modules is necessary to ensure 'freedom from interference' with other modules, as they serve as redundant safety systems, and act as the last safety instance, which can override the decision made in the higher level applications to avoid accidents or to minimize the impact of any accident when it is already unavoidable.

Besides the safety requirement, vehicle control modules are also responsible for the common lateral and longitudinal vehicle control capabilities an SDV should have. Here we introduce some of the simpler control capabilities needed for SDVs.

4.1.3.1   Lane keeping

The aim of the lane keeping function is to ensure that the vehicle remains within the current ego (own) lane boundaries. Usually, this means keeping the vehicle close to the middle of the lane. The key enabler for this function is obviously the lane boundary marking detection, as shown in Figure 4.7, which is normally performed by cameras. Many high-end vehicles already have similar functionality as part of their Advanced Driver Assistance Systems (ADAS) functions, which warn the driver when the vehicle is starting to inadvertently drift out of the current ego lane boundaries, e.g., without having the turn indicator explicitly switched on. The big difference is that the lane keeping in SDVs involves active lateral control of the vehicle, not merely a warning for the driver.

Figure 4.7: Lane (boundary marking) detection as a prerequisite for the lane keeping function. (Adapted from "Strada Provinciale BS 510 Sebina Orientale" by Grasso83. ©Grasso83, https://commons.wikimedia.org/wiki/ File:Strada_Provinciale_BS_510_Sebina_Orientale.jpg, "Strada Provinciale BS 510 Sebina Orientale", Coloration, https://creativecommons.org/licenses/by-sa/3.0/legalcode)

4.1.3.2   Adaptive cruise control

Adaptive Cruise Control (ACC), as shown in Figure 4.8, automatically adjusts the vehicle's velocity to keep a safe distance to the vehicle in front. ACC is also a well-known ADAS function and is an example of automatic longitudinal vehicle control. ACC works based on distance and velocity measurements of the other vehicle, typically performed by front radar sensors, front-facing cameras, or a combination of both. The performance requirement and test procedures for the ACC function are standardized in *ISO 15622* [11].

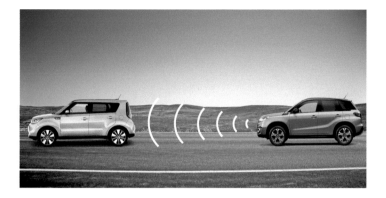

Figure 4.8: ACC based on radar distance and velocity measurement of the vehicle in front.

4.1.3.3   Lane changing

The automated lane changing function allows the vehicle to safely change from one lane to another. Lane changing, as shown in Figure 4.9, is more complex than the two previous functions, as it involves not only both lateral and longitudinal control, but also relies on the information from many sensors, as well as the reliability of other functions, such as object detection in neighboring lanes.

Figure 4.9: Automated lane changing. (Adapted from Pxhere. CC-Zero-1.0)

4.2 System architecture

The previous section introduced the basic capabilities or functions a typical SDV needs to master. However, from the software point of view, each of the functionalities can be seen as an individual feature to be implemented, maintained and improved. Building such a complex system as an SDV requires a huge amount of effort. SDV manufacturers typically do not have all the resources to build everything from scratch by themselves. Thus, SDV applications need to rely on software components from third parties which in turn have dependencies on other software. This section describes how the software components are typically needed for a complete SDV software stack and explains the relationship between the different software components in an SDV system architecture.

A good system architecture is essential when building any complex software product. Having a good system architecture not only minimizes the overall development and maintenance effort in the long run, but it also makes the system easier to improve or update in the future. A good system architecture generally combines widely used design patterns and well-proven design principles with innovative approaches to mitigate application-specific problems, ideally by taking advantage of the local characteristics of the system.

Vertical abstraction is commonly used as a way to simplify the architecture of complex software systems. The system is divided into several sub-systems, or layers. The implementation detail of each layer is mostly hidden from the other layers, and is only visible through a pre-defined interface. Each layer interacts only with its immediate higher or lower layer, hence the name multi-layered system architecture. Although the multi-layered system approach adds extra complexity compared to a monolithic system architecture, where each component can interact freely with any other components in the system, the advantages often amortize its cost for the long run, as it promotes better maintainability, implementation flexibility, and reusability.

Although there can be some variations, a typical SDV system architecture can be generalized to three layers: hardware, middleware, and application layer.

4.2.1   Hardware layer

The hardware layer is responsible for accessing and controlling the sensors, actuators and other hardware in the SDV. Depending on the interface used, the low-level hardware access is provided by some device-specific proprietary firmware or by some generic driver software that supports the necessary standard. Except where finer low-level control or more optimized performance using a hardware-specific Application Programmable Interface (API) is needed, the use of standardized hardware interfaces is desirable to ensure hardware-independency, to avoid vendor lock-in and to promote better code reuse. Almost all software in this layer is normally provided or made available by the hardware manufacturers. Depending on the level of complexity of the hardware, however, regular firmware update might be necessary to keep the hardware safe and secure for use.

4.2.2   Middleware layer

The middleware layer typically consists of an operating system and a runtime environment. It interfaces between the hardware layer below and the SDV application layer above. Using a middleware layer brings many benefits due to the availability of ready-to-use, well-tested, libraries for the application layer. These include reduced development effort, improved maintainability and platform independence. Additionally, popular middlewares also typically come with a suite of useful tools for application development and testing that can further reduce development effort. Software for the middleware layer typically comes either from open source projects or from software vendors. Some of the widely used middlewares for developing SDV will be introduced in the next section.

4.2.3   Application layer

All the features described in the functional architecture section above are implemented as software components in the application layer. Unlike the other two layers, which are normally provided as standard or off-the-shelf software products, the application layer contains tailor-made or highly customized software components that distinguish one SDV system from another. Most of the time, software in this layer interacts with other software components and the rest of the system through an API provided by the middleware. The interaction between the software components and middleware may be based on message passing, shared memory, function calls, etc. depending on the communication mechanism supported. Using an abstract interface like this allows the application layer to be middleware-independent.

4.3   SDV middleware examples

As implied in the previous section, choosing the right middleware is very important. In this section, we will look at three middleware stacks often used for SDV development.

4.3.1   Robot operating system

The name Robot Operating System is somewhat of a misnomer since ROS is not an operating system in the traditional sense of Windows, Linux or MacOS. Rather, it is a middleware that runs on top of an operating system. ROS was pioneered by Willow Garage, a Silicon Valley-based robotics lab, in 2006. ROS is a popular open source middleware that has been widely used by researchers and industry in the field of robotics, and autonomous driving.

As a middleware, it provides a collection of tools and libraries that facilitate the development of robot applications. As well as offering a communication infrastructure that supports seamless distributed communication across machines, it also supports different modes of communication such as asynchronous (using a topic), synchronous (using a service), and data storage (using a parameter server) [3]. Thanks to its flexible client library architecture, ROS applications can be implemented using many programming languages, including C/C++, Python, C#, Ruby, Go and Java. However, support for languages other than C/C++ and Python is still considered experimental at the time of writing.

As shown in Figure 4.10, the ROS framework is made up of a ROS master node and a number of user nodes that receive input and transmit output data using ROS messages. The ROS master node is the central registration point of all nodes, and also provides the parameter server, i.e., a special service for accessing parameters at runtime. During initialization, a node registers itself with the master node, and informs the master node about the topic or any services the node is interested in or can provide. Or in ROS jargon, the topic it wants to subscribe to, and the service or topic it will publish at runtime. As soon as a new message is available, that message will be sent directly from the publisher to all registered subscribers by means of a peer-to-peer Remote Procedure Call (RPC). The direct peer-to-peer communication architecture allows efficient and scalable communication, as opposed to a centralized communication, where all messages are routed centrally through the ROS master node before they finally get distributed to their subscribers. A node can also register services that can be invoked by other nodes in a request/response manner.

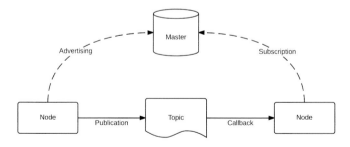

Figure 4.10: ROS framework overview. (©Noel.martignoni, https://commons.wikimedia.org/wiki/File:ROS-master-node-topic.png, https://creativecommons.org/licenses/by-sa/4.0/legalcode)

To promote code reusability and collaborative development in large systems, ROS software is organized in packages, and sometimes also stacks. An ROS package is the atomic build unit or module to solve a particular task. It is simply a directory that contains the ROS nodes, some libraries, a CMake build file and an XML package metadata that describes the package name, version, dependencies, etc. An ROS stack is a collection of ROS packages that collectively provide a certain functionality, such as navigation.

SDVs, like other robotics systems, typically work with various coordinate systems, e.g., the world coordinate system from GNSS sensors, a coordinate system relative to the vehicle's center of gravity, one relative to some fixed reference frame for localization, etc. Keeping track of the coordinate frames, and calculating the transformation from one system to another, is a tedious and error-prone task. The ROS transform system (*tf*) is a coordinate frame tracking system using a standardized protocol for publishing transforms, which is designed to simplify this task. Using tf, ROS nodes can publish coordinate frames using their 'native' coordinate systems, or retrieve frames in their preferred coordinate systems, without having to know all the coordinate frames in the system. Internally, the tf system maintains the hierarchical relationship between coordinate systems using a tree data structure. The necessary transformation between source and target frame is obtained by calculating the net transform of the spanning set formed by walking up the edges of the tree until a common parent node is found [7].

Due to its open-source nature, and its active development within the robotics community, ROS has become a very popular and widely used middleware for research and development of robot applications, including SDVs. One of the key advantages of using such a popular framework is the availability of various helpful publicly available open source ROS packages or libraries, which can be used as the base for creating customized solutions, for rapid prototyping, or simply as a good resource for learning robotics algorithms.

4.3.2 *Automotive data and time-triggered framework*

Automotive data and time-triggered framework (ADTF) was initially developed by Audi Electronic Venture (AEV) as an internal framework for development and test of Advanced Driver Assistance Systems (ADAS), before it was made available as proprietary software for other car manufacturers and automotive suppliers in 2008. Because of its strong automotive background, ADTF has robust and extensive support for automotive-specific devices and interfaces, e.g., support for LIN, CAN,

MOST and FlexRay buses and is widely used, especially in the German automotive community. Similar to ROS, ADTF is a middleware that runs on top of other operating systems, such as Linux or Windows.

As shown in Figure 4.11, ADTF is a multi-layered framework, with four software layers: the component, the launcher environment, the system service, and the runtime layer. The component layer comprises user custom filters, as well as toolboxes that contain additional filters or services, e.g., for interfacing with I/O hardware either provided by ADTF or by third-party suppliers. The filters are loaded and run in one of the launcher environments, including the headless (non-GUI) console environment, a minimalistic graphical user interface (GUI) runtime environment, or a full-blown GUI development environment that includes a configuration editor, profiler, and debugger. The system service layer provides the necessary functionality for the launcher execution, as well as other basic services, such as memory pool and clock. Finally, the runtime layer is responsible for the registering of components, scheduling of system services and for changing the system runtime level [5].

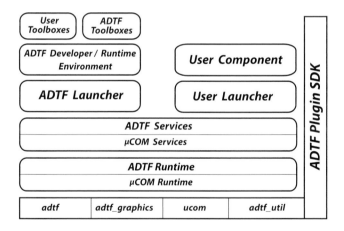

Figure 4.11: ADTF architecture overview. (Adapted from "ADTF Architecture Overview" by Digitalwerk, 2019, ADTF v2.14.3 Documentation. ©2019 Digitalwerk GmbH)

ADTF applications are generally implemented as filters and services, written in Python or C/C++ using the ADTF Software Development Kit (SDK). In ADTF, existing functionalities in the application layer are typically easily wrapped as filters and the input/output of the functions are stimulated through the pins. The creation of data visualizations or a Graphical User Interface (GUI) is supported through the integrated Qt graphical framework. The use of platform-independent ADTF and Qt SDKs ensures that the same application code can run on both Linux and Windows versions of ADTF. Similar to ROS, ADTF also supports built-in tools and functionalities for data recording and playback that are useful for offline processing. The data exchanged between filters ranges from primitive data types, such as int and bool, to complex user data types, such as ones represented using hierarchical structured data (nested structs) and dynamic arrays. In order for the receiver to correctly interpret the incoming data stream, a description of the transmitted media type, written in the ADTF Data Definition Language (DDL), needs to be made available to the receiving party at compile-time or exchanged dynamically as part of the data stream at run-time.

The new version of ADTF, ADTF 3, allows seamless communication between ADTF instances across distributed systems. Unlike previous versions, each ADTF instance is started as a separate process instead of a single thread within the ADTF runtime/development environment process, and the communication between multiple instances is realized through either using the host system's Inter Process Communication (IPC), or using common network communication protocols, such as Transmission Communication Protocol (TCP), User Datagram Protocol (UDP), or Stream Control Transmission Protocol (SCTP). Another noteworthy improvement is support for Modern C++, which enables developers to write their ADTF code in a more clear and concise way. The built-in support for generic Remote Procedure Call (RPC) communication with other filters enables ADTF developers to develop distributed control-flow-based applications in a more straightforward way, in addition to the classical filtering or data-flow-based applications.

4.3.3 Automotive open system architecture

Automotive open system architecture (AUTOSAR), unlike ROS and ADTF, is actually a set of standards, rather than a middleware software per se. The standards are published by the AUTOSAR consortium, a global development partnership of companies within the automotive ecosystem, to create a standardized architecture for automotive electronic control units (ECUs). Thus, AUTOSAR middleware simply means middleware that complies with the AUTOSAR standards, and might come as a collection of software products from various companies, each being specialized in implementing some or all of the standards.

Before AUTOSAR, every car manufacturer had to develop their own proprietary systems or use proprietary systems from their suppliers. The lack of standards resulted in poor code reusability, lack of systems interoperability, limited testability (which leads in turn to poor software quality) and high development and maintenance costs in general. Not only does AUTOSAR offer economic benefits for players in the automotive ecosystem, but its methodology and specifications support the development of safety-critical automotive applications in accordance to *ISO 26262*.

In the AUTOSAR architecture, the software that runs on the automotive ECU is divided into three abstraction layers: application, runtime environment (RTE) and the basic software, as shown in Figure 4.12. The application layer contains user or application specific software components (SWC). The communication between SWCs, or between an SWC and a communication bus or other service is done through the RTE. The RTE is actually machine-generated code that implements the communication 'plumbing' between those components, i.e., creating the necessary internal variables to store the exchanged data and methods for the SWCs to access those variables or AUTOSAR services. The basic software layer provides hardware abstraction and standardized services, e.g., diagnostics, coding, etc. for the SWCs, and RTE to perform their duties. The basic software layer stack, along with the RTE generator and other tools, e.g., AUTOSAR authoring/modelling IDE, are typically provided by AUTOSAR technology supplier companies.

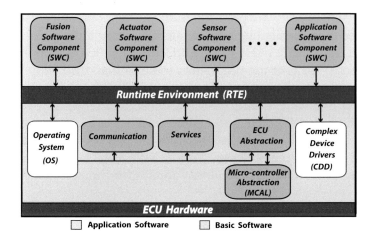

Figure 4.12: AUTOSAR architecture. (Adapted from "Layered Software Architecture", by AUTOSAR, 2017, AUTOSAR Classic Platform 4.3.1 Document ID 053, p. 80. ©2017 AUTOSAR)

The communication between components in AUTOSAR falls into two general paradigms: *Sender/Receiver (S/R)* and *Client/Server (C/S)* communication. A typical application involves reading input values from the communication bus, e.g., a sensor data stream, processing the input values in the SWC and sending the computation result back to a communication bus periodically. This kind of cyclic data-flow-based processing is mostly realized using an S/R interface with some additional parameters, such as whether the data is queued on the sender/receiver port, blocking/non-blocking access, etc. Control-flow-based communication or non-periodic access to services, such as reading a certain vehicle parameter (coding) or a value-returning function call to cryptographic service, are typically handled using C/S communication.

Regardless of the type of communication used, the type and format of data to be exchanged between the participants of the communication needs to be agreed upon and specified at design time. The method signature, which includes the method name, the expected parameters and their types, of all available functions accessible in the SWC as well as the method signature of required functions from outside the SWC are the most important parts of the specification. Because the specification is usually exported from AUTOSAR XML or ARXML into C header files and serves a 'binding contract' to all components that interact with the SWC, this is known as the *contract phase header*. ARXML is the common data exchange format for AUTOSAR. It is a human-readable format (XML), and is used by all tools that work with AUTOSAR.

The SWC runtime behavior is also defined at design time. The definition specifies how the SWCs get scheduled, i.e., the cycle parameter if the SWC needs to be scheduled periodically, or the event types for event-based invocation. The generated RTE also contains a scheduler that allows all SWCs to be scheduled. In the AUTOSAR composition, all software components in the ECU are 'tied' together, i.e., all sender/server ports are bound to their listener/client counterparts. The SWCs are eventually integrated together along with the generated RTE, and other components during the linking step. After the linking is successful, the resulting binary is flashed to the ECU, and is ready for execution on the vehicle.

Because AUTOSAR is conceptually designed to enable a standardized approach for developing safe hard real-time automotive applications, everything, or almost everything, has to be statically configured at design time. The static system approach certainly ensures the application is fully deterministic. However, it might be too restrictive for certain applications, such as SDV. Therefore, the AUTOSAR consortium introduced the *AUTOSAR adaptive platform standard* with some relaxed constraints, to enable the development of dynamically configurable systems, such as support for execution from Random Access Memory (RAM) instead of Read-Only Memory (ROM), dynamic scheduling, and virtual address space [8].

At the time of writing, AUTOSAR has published the following standards [1]:

■ *Classic Platform Standard* for 'traditional' hard real-time and safety-critical systems

■ *Adaptive Platform Standard* for systems that require dynamically linked services and clients

■ *Foundation Standard* that provides the common parts of the classic and adaptive platforms

■ *Acceptance Tests Standard* for validation of AUTOSAR stack implementations at bus and application level

■ *Application Interfaces Standard* that defines the syntax and semantics of common domain application interfaces

4.4 Summary

As we have seen in this chapter, SDVs require the coordination of a large number of discrete functions in order to achieve the aim of safely driving to their destination. In this chapter we showed how the vehicle can combine its knowledge of the local environment with other factors in order to reach its destination. The first section explored the additional functionality that an SDV needs in order to work safely and effectively. These functions are divided into planning and vehicle control.

The planning task answers the question, 'How do I safely reach my destination?'. It is split into three distinct phases: route planning, behavioral planning and motion planning. Route planning aims to find the optimal route through the map between the current location and the destination. Optimal means this task needs to take account of all constraints from the occupant (e.g., a preference to avoid tolls), the vehicle (e.g., conserve battery life in an electric vehicle), and the environment (e.g., do not drive the wrong way along a freeway). The result of the route planning is a series of discrete waypoints. Behavioral planning seeks to find the optimal way to reach the next waypoint, taking account of the current surroundings and road conditions. Finally, motion planning is about determining the optimal path to achieve the output of the previous step, and providing the high level list of commands that the vehicle controller will use to drive the vehicle.

The vehicle control functions pass instructions to the assorted actuators and controllers that actually drive and steer the vehicle. It takes the list of steering angle, throttle, and brake commands from the planning function, and ensures these are safely translated into commands to pass to the hardware. As we explained, the vehicle controller is a safety-critical system, and is the 'last line of defense' if any unexpected event occurs (e.g., due to external factors or sensor failures). As we saw, there are three main vehicle control tasks: lane keeping, adaptive cruise control and lane changing. All three tasks exist as part of the Advanced Driver Assistance Systems (ADAS) that are fitted to many modern vehicles. However, as was explained, in an SDV they have to perform all functionalities without human intervention. This is especially challenging for safe lane changing where the SDV has to perform several control tasks at once as well as monitoring a large number of sensors.

The second part of the chapter introduced the multi-layered system architecture abstraction. The multi-layered system architecture describes the actual software system that achieves the functions needed for the SDV. As we saw, it is divided into three layers. The hardware layer on which the system runs, the middleware layer that passes data from the hardware up to the application layer and passes commands back down to the hardware and finally the application layer which implements the detailed functions described above. As you will recall, in general, the hardware and middleware layers utilize commodity software provided by hardware vendors, or from open source and third party libraries. This allows a far greater degree of interoperability and simplifies both development and maintenance of the code. The application layer by contrast tends to consist of customized or proprietary code optimized for each SDV.

The final part of the chapter looked in more detail at some standard middlewares used for SDV applications. The middleware is a critical part of any SDV software system. As we saw, the Robot Operating System (ROS) is a popular choice for SDVs because it is widely researched and has a large library of applications and tools. The Automotive Data and Time-Triggered Framework (ADTF) is a proprietary middleware specifically created by Audi for automotive applications. As a result, it offers strong support for the proprietary hardware and control buses found in vehicles. Unlike ROS and ADTF, AUTOSAR (Automotive Open System Architecture) is a set of standards that define the requirements for any automotive electronic control unit. By following the AUTOSAR standards, SDV manufacturers can ensure they are creating a safe, maintainable, and interoperable middleware.

Now that we have an idea of what hardware and software is needed for an SDV, the next step is to combine these into a complete system. This is the bulk of any developer's work when creating an SDV, and it requires detailed knowledge of coding. As with most technology, there are a wide range of choices for both the hardware and software. However, there are several popular frameworks which we will discuss in the next chapter.

References

[1] AUTOSAR. Autosar standards. https://www.autosar.org/ standards/. [accessed 07-Nov-2018].

[2] Laurène Claussmann, Marc Revilloud, Sébastien Glaser, and Dominique Gruyer. A study on al-based approaches for high-level decision making in highway autonomous driving. In *Systems, Man, and Cybernetics (SMC), 2017 IEEE International Conference on*, pages 3671–3676. IEEE, 2017.

[3] K Conley. Ros/introduction-ros wiki. *ROS Wiki*, 2011. http://wiki. ros.org/ROS/Introduction. [accessed 07-Nov-2018].

[4] Julian Dibbelt, Ben Strasser, and Dorothea Wagner. Customizable contraction hierarchies. *Journal of Experimental Algorithmics (JEA)*, 21:1–5, 2016.

[5] Digitalwerk. Automotive data and time-triggered framework sdk documentation overview version 2.14.2. https://support. digitalwerk.net/adtf/v2/adtf_sdk_html_docs/index.html. [accessed 07-Nov-2018].

[6] Edsger W Dijkstra. A note on two problems in connexion with graphs. *Numerische mathematik*, 1(1):269–271, 1959.

[7] Tully Foote. tf: The transform library. In *Technologies for Practical Robot Applications (TePRA), 2013 IEEE International Conference on*, pages 1–6. IEEE, 2013.

[8] Simon Fürst. Autosar adaptive platform for connected and autonomous vehicles. In *Proc. conf., 8th Vector Congress, Alte Stuttgarter Reithalle*, 2016.

[9] Robert Geisberger, Peter Sanders, Dominik Schultes, and Christian Vetter. Exact routing in large road networks using contraction hierarchies. *Transportation Science*, 46(3):388–404, 2012.

[10] GraphHopper. Graphhopper routing engine. `https://github.com/graphhopper/graphhopper`. [accessed 07-Nov-2018].

[11] ISO. Iso 15622: Intelligent transport systems - adaptive cruise control systems - performance requirements and test procedures. `https://www.iso.org/standard/71515.html`. [accessed 07-Nov-2018].

[12] Christos Katrakazas, Mohammed Quddus, Wen-Hua Chen, and Lipika Deka. Real-time motion planning methods for autonomous on-road driving: State-of-the-art and future research directions. *Transportation Research Part C: Emerging Technologies*, 60:416–442, 2015.

[13] Lydia Kavraki, Petr Svestka, and Mark H Overmars. *Probabilistic roadmaps for path planning in high-dimensional configuration spaces*, volume 1994. Unknown Publisher, 1994.

[14] Kavraki-Lab. The open motion planning library. `http://ompl.kavrakilab.org`. [accessed 07-Nov-2018].

[15] Steven M. Lavalle. Rapidly-exploring random trees: A new tool for path planning. Technical report, 1998.

[16] Alaa Sheta and Hamza Turabieh. A comparison between genetic algorithms and sequential quadratic programming in solving constrained optimization problems. *ICGST International Journal on Artificial Intelligence and Machine Learning (AIML)*, 6(1):67–74, 2006.

[17] Junqing Wei, Jarrod M Snider, Tianyu Gu, John M Dolan, and Bakhtiar Litkouhi. A behavioral planning framework for autonomous driving. In *2014 IEEE Intelligent Vehicles Symposium Proceedings*, pages 458–464. IEEE, 2014.

Chapter 5

Putting it all together

The previous chapters introduced the hardware and software components used in autonomous vehicles. The next step is to describe how we put these components together to build a self-driving vehicle (SDV). This chapter covers three main topics: the steps required to prepare for SDV development, some walk-through examples of processes such as installing sensor drivers and reading vehicle data, and a discussion of SDV testing.

This chapter includes a discussion of Open Source Car Control (OSCC), a by-wire control kit designed to enable computer control of a modern vehicle. OSCC allows users to connect to a vehicle's internal control system, including its communications network. This opens up the option of developers using their own hardware and software, in conjunction with Arduino-based OSCC modules, to send control commands to vehicle component ECUs, and to read control messages from the vehicle's CAN network. The OSCC kit offers tremendous potential for independent developers working on a tight budget who are unwilling to spend large sums of money on proprietary, black-boxed systems.

Please remember the disclaimers in the introduction to this book. Modern vehicles are extremely complex and contain components that can kill or severely injure you if handled incorrectly. At the least, modifying your vehicle is likely to void any warranties, and may void your insurance. Furthermore, in most jurisdictions, SDV development on public roads can only be done with a special license, and in some jurisdictions SDVs are illegal. It is particularly important to realize that OSCC is solely designed to facilitate prototyping in controlled, private environments. OSCC modules do not meet the required safety or quality standards for use on public roads.

5.1 Preparation

Before you can start to develop an SDV, there are some preparatory steps you need to complete, such as setting up the donor vehicle and installing and calibrating all the sensors.

5.1.1 Choosing your vehicle

Choosing a suitable vehicle is key to SDV development. As mentioned in Chapter 2, modern vehicles often have proprietary actuators, and ECUs that are hard to access. This can mean an older vehicle makes a better donor. However, older vehicles lack many of the required by-wire control systems. This is where something like OSCC comes in. OSCC has been developed to work with a specific, readily available, donor vehicle. This makes development easier. We will discuss OSCC in detail in Section 5.2.1.

5.1.2 Vehicle network

Having chosen your vehicle, you need a way to seamlessly integrate the vehicle's hardware and software. Only by building robust hardware connections can you create a stable network through which the sensors, computing platform, and actuators can communicate with each other. Ultimately, we need the vehicle components to behave exactly as expected in all cases. Any glitch in communication could lead to errors, and prevent the system from functioning effectively, and safely. All modern vehicles have some form of integrated network, and many may come with more than one. Given the differing needs of different types of sensors, most SDVs will need some combination of different networks.

In Chapter 2 we discussed the relative performance of the different network types. Here we will look at how you might decide between these. One of the primary considerations is the available bandwidth or network capacity. Some sensors, such as lidar, will generate enormous amounts of data that need to be fed back to the computing platform. Other sensors, such as odometers, have much lower data rates. Another consideration is whether the data processing needs to be real-time or not.

Safety critical systems that are trying to detect and avoid collisions have real-time constraints, while systems that are feeding the route planning algorithm may be able to cope with some delay (though even here you want to minimize delays). Finally, cost and practicality aspects ought to be considered as well.

Depending on the vehicle platform, it may also be necessary to connect the communication networks to the *vehicle's central gateway* module. This gateway acts as a router, providing a central hub for all vehicle communication. Information from each component of the vehicle travels to the gateway, which exchanges data between the buses of whichever networks are present in the vehicle. This solution allows the computing platform to access data from other, non-driving components of the vehicle, such as the vehicle's temperature and battery status.

5.1.3   Sensor selection and calibration

Choosing the right sensors for your vehicle is a balance of three things:

- Functionality
  The specific use case for your SDV may dictate which sensors you need. For instance, an automated forklift in an indoor warehouse will need a very different set of sensors than a passenger vehicle designed to work on freeways, and in town.

- Cost
  Some sensors, such as lidar, offer excellent performance, but can be prohibitively expensive. Sometimes, it is cheaper to combine more affordable systems such as radars and cameras and use sensor fusion to synthesize the data you need.

- Practicality
  You need to consider a number of more practical aspects too. These include the available power in the vehicle, where (and how) to mount the sensors, and how to connect the sensors to the in-vehicle network.

It is essential that the information collected and delivered by SDV sensors is as accurate as possible. One way we achieve that is by calibrating them. In the case of cameras, for example, calibration makes image interpretation more accurate, enabling precise calculations of distance and speed. The calibration process determines a sensor's precise location relative to the vehicle and its surroundings and ascertains whether it differs

Figure 5.1: Example of how sensor calibration is performed. (Reprinted with permision from STR Service Centre. ©STR Service Centre Ltd.)

from the default parameters. Sensor position, orientation and scaling are some examples of parameters that may differ from vehicle to vehicle.

Typically, cameras are calibrated by placing a special calibration target at a pre-defined distance from the camera as shown in Figure 5.1. This distance is specified by the manufacturer. The device compares the image with known values, calculates offsets, and makes the necessary adjustments relative to the initial parameters to bring the values into line.

Calibration is not only required during initial set-up, however; adjustments need to be made to the sensors in the vehicle from time to time to reflect changing conditions. This is typically performed using self-calibration algorithms in the software. According to Collado et al. [4], it is possible to extract road markings and use them for calibration of stereo cameras using an algorithm that finds the 'height, pitch, and roll parameters of the vision system.'

5.2 Development

Once you have chosen a suitable donor vehicle, selected and installed your sensors, installed the computing platform and ensured all the hardware is networked together and installed in the vehicle properly as illustrated in Figure 5.2, you are ready to start development. In this section we

Figure 5.2: Example of a development and measurement system inside an SDV prototype car. (©Steve Jurvetson, https://commons.wikimedia.org/wiki/File: Robocar_3.0.jpg), "Robocar 3.0", https://creativecommons.org/licenses/by/2.0/ legalcode)

will give some practical examples of the steps needed to install the middleware, access the sensors and start actually developing the software.

The computing platform that unifies the entire autonomous vehicle needs to run an appropriate middleware. Various options are available, but in this case we will base our examples on the open-source middleware Robot Operating System (ROS), which we discussed in the previous chapter. This collection of software frameworks is an ideal choice because it is designed for rapid prototyping. However, due to a lack of real-time behavior, it is not suitable for series production, though efforts are already underway to resolve this and other issues by creating next-generation ROS architectures such as ROS 2.0 [7] and RGMP-ROS [11].

5.2.1 Open source car control

Open Source Car Control (OSCC) is a fully open source vehicle control project which intends to open up SDV development to a larger audience. To simplify the SDV development process, OSCC focuses on a single, readily available, vehicle model, namely the Kia Soul. OSCC enables engineers to construct their own SDV applications by providing the required hardware and software for manipulating and modifying the by-wire systems of 2014 and later models of this car. This offers a major benefit to SDV developers, who would otherwise have to work from scratch in figuring out how to manipulate the vehicle. It also works out to be far cheaper. Buying the required components for an OSCC project can cost less than USD $10,000 (including the vehicle platform). By contrast, purchasing the components and vehicle for a bespoke by-wire vehicle platform might cost over USD $100,000. What's more, the information contained in some by-wire vehicle platforms is 'black-boxed', which means that the information it contains is inaccessible beyond a certain point, in order to protect proprietary technology. OSCC enables anyone interested in this field to experiment with manipulating by-wire systems, because the required black box data has already been analyzed, and made publicly available.

The OSCC project consists of three main components. Hardware controllers, based on the *Arduino* microcontroller platform, the OSCC controller software, and a detailed wiki or knowledge base giving details of things like hardware designs, CAN-bus control codes, etc.

5.2.1.1   OSCC controllers

The OSCC project tries to re-use the Kia Soul's original hardware where possible. However, there are several additional modules that are needed in order to convert it into a full SDV.

The OSCC project makes extensive use of the Arduino family of open-source microcontroller boards. Arduino boards typically consist of a microcontroller (often an Atmega chip), some memory, a stable power supply circuit, and a number of digital and analog pinouts accessible via a standard set of headers. These headers allow you to attach a large number of 'shields', which are daughter boards providing specialized functions, such as networking, servo controllers or sensor inputs.

CAN bus gateway

The Kia Soul already has several CAN buses installed which are able to provide data such as steering angle, brake pressure, wheel speeds and turn indicator. However, in order to avoid interference with the native systems, the OSCC system uses a CAN bus gateway to connect to the car's OBD-II CAN bus. The gateway then allows messages from the OBD-II CAN bus to be shared with the new OSCC control CAN bus. The gateway is based on the Seeed Studio CAN-BUS Shield v 1.2.

OSCC controller boards

The system needs several controller boards. Specifically, three modules for connecting with the by-wire systems, the CAN gateway described above, and a power distribution board that includes an emergency-stop capability. All these modules are Arduino shields, with all the necessary schematics provided on the OSCC wiki. Custom fabrication shops should be able to produce the boards for about USD $50 each.

5.2.1.2   X-by-wire systems

The Kia Soul is shipped with full actuator-controlled steer-by-wire and throttle-by-wire. However, braking is still purely mechanical. The OSCC project team recommend retrofitting the brake-by-wire system from a 2004-2009 Toyota Prius. The OSCC Wiki gives full details on how to do this, including diagrams showing how it is installed in the Prius, along with pinouts for connecting it to the OSCC controller.

5.2.1.3   OSCC software

OSCC software consists of the necessary firmware for all the Arduino boards. Firmware is the embedded software that controls the hardware components. There are also some elements of control software that are designed to work with a standard SDV middleware, such as ROS. The software and firmware is based on PolySync's Core platform, and full instructions for building and installing are provided on the OSCC wiki. Below, we give an example of how to build the firmware for a petrol Kia Soul.

```
# navigate to the correct directory
cd firmware
mkdir build
cd build

# use flags to tell cmake to build for a petrol vehicle and
    to override operator control
cmake .. -DKIA_SOUL=ON -DSTEERING_OVERRIDE=OFF

# now build the firmware with make
make

# alternatively, use make <module-name> to build a specific
    module
make brake
make can-gateway
make steering
```

```
make throttle

# now you can upload each module. The system expects a single
    module connected to /dev/ttyACM0
make brake-upload

# if you want to upload all the modules at once specify their
    addresses
cmake .. -DKIA_SOUL=ON -DSERIAL_PORT_BRAKE=/dev/ttyACM0
    -DSERIAL_PORT_CAN_GATEWAY=/dev/ttyACM1
    -DSERIAL_PORT_STEERING=/dev/ttyACM2
    -DSERIAL_PORT_THROTTLE=/dev/ttyACM3

# now you can flash all the modules at once
make all-upload
```

OSCC API

OSCC includes an API for controlling the modules, and for reading in sensor values from, e.g., the steering system. Here are some code snippets for accessing the API.

```c
// open the OSCC endpoint
oscc_result_t oscc_open( uint channel );

// enable all modules
oscc_result_t oscc_enable( void );

// publish control commands to the relevant CAN buses
oscc_result_t publish_brake_position( double
    normalized_position );
oscc_result_t publish_steering_torque( double
    normalized_torque );
oscc_result_t publish_throttle_position( double
    normalized_position );

// subscribe to the relevant callbacks to receive sensor data
oscc_result_t subscribe_to_brake_reports(
    void(*callback)(oscc_brake_report_s *report) );
oscc_result_t subscribe_to_steering_reports(
    void(*callback)(oscc_steering_report_s *report) );
oscc_result_t subscribe_to_throttle_reports(
    void(*callback)(oscc_throttle_report_s *report) );
oscc_result_t subscribe_to_fault_reports(
    void(*callback)(oscc_fault_report_s *report) );
oscc_result_t subscribe_to_obd_messages(
    void(*callback)(struct can_frame *frame) );
```

```
// close the OSCC endpoint
oscc_result_t oscc_close( uint channel );
```

5.2.2 Installing middleware and device drivers

Once you have wired all the components of the vehicle together, you need to install and configure them correctly to ensure they work as intended. Each device should come with the correct device drivers. These must be installed to ensure that the device can communicate successfully with the middleware and the rest of the vehicle.

5.2.2.1 ROS

As explained in Section 4.3.1, Robot Operating System (ROS) is a middleware that runs on top of another operating system such as Linux. As a middleware, it provides a collection of tools and libraries that facilitate the development of robot applications. As well as offering a communication infrastructure that supports seamless distributed communication across machines, it also supports different communication modes: asynchronous (using a topic), synchronous (using a service), and data storage (using a parameter server).

ROS is an open-source program, so it is freely accessible to anyone. The easiest option is to download ROS[1] and simply follow the installation instructions. There are several versions of ROS, ranging from stable versions with a long development history to more recent versions that have less robust support. In the following examples, we have used the *indigo* version.

5.2.2.2 Sensor drivers

As mentioned above, each hardware component comes with firmware or a driver that must be installed before the device can be used in the vehicle. In this example, we will be working through the installation process for a Velodyne VLP 16 lidar sensor. Let us take a look at the steps required to connect the sensor to the middleware, highlighting the key installation steps and command line inputs that make it all happen.

Before using the Velodyne lidar, you need to install the driver:

```
$ sudo apt−get install ros−indigo−velodyne
```

---

[1]http://www.ros.org/ROS/installation

Alternatively, you can build the driver from the source yourself as described in [9]. First, you need to initialize the *ROS indigo* environment:

```
$ source /opt/ros/indigo/setup.bash
```

The next step is to clone (download) the driver's source code in the ROS workspace:

```
$ mkdir −p ~/catkin_ws/src
$ cd ~/catkin_ws/src
$ git clone https://github.com/ros−drivers/velodyne.git
```

Now you can install the driver's dependencies:

```
$ cd ..
$ rosdep install —from−paths src —ignore−src —rosdistro
    indigo −y
```

Finally, you can build the driver and initialize it in the ROS workspace:

```
$ catkin_make
$ source devel/setup.bash
```

5.2.2.3 CAN driver

Vehicle components can communicate over a number of different networks. In our example, the vehicle platform and central gateway will communicate by exchanging CAN messages, using a *SocketCAN* driver which has been included in the Linux kernel since version 2.6.25. To use the driver, we start by loading it using the following commands:

```
$ sudo modprobe can
```

The lsmod command tells you whether the *SocketCAN* module has loaded successfully:

```
$ sudo lsmod | grep can
can  45056  0
```

This tells us that the *SocketCAN* driver has been loaded successfully. The number 45056 simply indicates the amount of memory occupied by the driver in bytes, while the last number tells us how many instances of the module are currently being used. Zero means that no instances are in use. The next step is to configure the CAN bit rate. Devices operating on a CAN network must all operate on the same bit rate. The following example shows how to set the bit-rate to 1,250,000 bps for the first CAN interface (labeled can0). There may be more than one interface available for the computing system. These are then initialized as can1, can2, and so on.

```
$ sudo ip link set can0 type can bitrate 1250000
```

Finally, we can start the driver:

```
$ sudo ifconfig can0 up
```

Some computing platforms do not have a CAN interface output, in which case you need to install a USB to CAN hardware adapter. In our example we are using ROS as a middleware, which gives us the option of installing an extremely useful ROS package called socketcan_bridge. This package allows CAN messages to be translated to ROS topics, and vice versa. Here's how to install socketcan_bridge as a package in ROS *indigo*:

```
$ sudo apt−get install ros−indigo−socketcan−bridge
```

5.2.3   Implementing the software

Next, we will briefly explain how to implement the software required to support SDV functionality.

Generally speaking, there are two main approaches you can take to implementing software: hand coding or model-based development. In practice, these are often used in tandem to reap the benefits of both approaches.

The steady advancement of artificial intelligence (AI), especially in the realm of deep learning, has prompted scientists to experiment with another approach, which allows computers to generate the 'rules' themselves, and then apply these rules to make decisions based on current input data. In this case, the functionality of the software is based on a decision-making process that is created entirely by computers, and that is not generally transparent to humans. We will discuss this topic in more detail in Chapter 6.

5.2.3.1   Hand-coded development

As the name implies, the approach involves manually writing the software's source code in programming languages such as C/C++. Some middlewares, such as Automotive Data and Time-Triggered Framework (ADTF) and ROS, also allow the software to be written in Python.

This approach gives software developers the most freedom and flexibility to implement the required functionality in software while taking into account general coding best practices. However, when it comes

Figure 5.3: Eclipse IDE with C/C++ Development Tooling Plugin.

to building safety-critical software, developers must also comply with common industry best practices, and formal sets of software development guidelines. Two examples of established guidelines in the automotive industry are MISRA-C, and MISRA-C++. As the names suggest, the Motor Industry Software Reliability Association (MISRA) [8] has defined these for C and C++, respectively.

Static code analysis tools, such as Helix QAC from Perforce Software or TestBed from LDRA, analyze the source code without actually running it (hence the term 'static') and report any coding standards, e.g., MISRA-C or MISRA-C++, compliance issues.

Source code can, of course, be written with any text editor. However, using an Integrated Development Editor (IDE) such as Eclipse (see Figure 5.3) may increase productivity due to features such as code completion, refactoring tools, and integrated debugging. Additional plugins such as PyDev and C/C++ Development Tooling (CDT) can also be installed to facilitate development tasks in Python and C/C++, respectively.

5.2.3.2 Model-based development

Model-based development uses visual modeling principles to specify and design software functionality. More complex functions are implemented as compositions of simpler function blocks, and the models can be automatically transformed into source code (typically in C) with the aid of a code generation tool. This means the models you create can be used to implement identical functionality across multiple computing

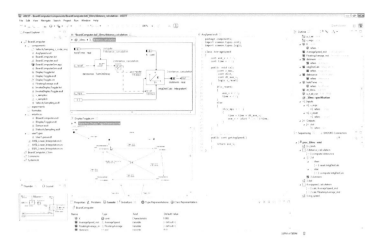

Figure 5.4: Model-based software development using ASCET. (Reprinted with permission from ETAS GmbH. ©2016 ETAS GmbH)

platforms, without having to deal with the intricacies of implementation or platform incompatibility issues.

Popular model-based development tools, such as MATLAB and ASCET (see Figure 5.4), provide extensive support for simulation as well as automatic model testing and verification, resulting in shorter development cycles [1]. Model-based development is a good choice of method for developing 'evolutionary' software applications, a process that generally involves creating new functionalities on top of the functionality of a previous version using iterated integration [3].

There are also some disadvantages to model-based development, however. Reportedly, these include lower coding efficiency, poor readability of the generated code, and the risk of vendor lock-in (i.e., difficulty in switching tool provider) due to a lack of standards [2]. What's more, some functionalities are much harder and more complex to express as a model than in code, such as those that involve recursive data structures [12].

5.2.4   Map building and localization

Access to accurate maps is essential for self-driving vehicles to localize themselves. Theoretically, you can achieve very precise localization if you have highly accurate maps, and robust localization technology, but, in some instances, these maps may have inadequate resolution or coverage, or even be completely unavailable. This problem is exacerbated by the fact that preliminary SDV testing is usually carried out at private facilities. Conducting such tests on public roads generally requires a special license from the local transportation authorities, such as the State of California's Department of Motor Vehicles [5].

If no sufficiently accurate map is available, then we need to build one ourselves. This typically involves traveling along the route while recording sensor data and running a Simultaneous Localization and Mapping (SLAM) algorithm on that live data stream, be it a video stream from the cameras or a point cloud from a lidar, as shown in Figure 5.5(a) and Figure 5.5(b). This task can be quite daunting as it takes a tremendous amount of effort to obtain and maintain these maps. The larger the area, the more effort is required. It is therefore worth considering the use of a map service or product from a specialized map company, such as HERE [6] or TomTom [10].

Once the map has been built, the localization algorithm processes data from the sensors to determine the vehicle's most likely position and pose on the map. As the vehicle moves on and gathers more information from other input data (sensor fusion), the system automatically makes any necessary adjustments to the results of the localization algorithm.

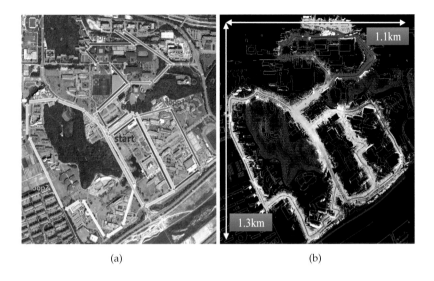

(a)   (b)

Figure 5.5: (a) Example of a route with multiple loops for building a 3D map. (b) The generated 3D map with point cloud. (Adapted from "Accurate Mobile Urban Mapping via Digital Map-Based SLAM", by Hyunchul Roh, Jinyong Jeong, Younggun Cho, Ayoung Kim, 2016, Sensors 2016, 16(8):1315. ©Hyunchul Roh, Jinyong Jeong, Younggun Cho, Ayoung Kim, https://www.mdpi.com/sensors/sensors-16-01315/article_deploy/html/images/sensors-16-01315-g012. png, "3D mapping result and data logging path", Ordering, https://creativecommons.org/licenses/by/4.0/legalcode)

5.2.5   Reading vehicle data

The data provided by the sensors is raw and needs to be interpreted. For example, we need to capture and interpret the data from a vehicle's turn signals. According to the OSCC project, in a Kia Soul, the turn signals are determined by evaluating byte 5 of CAN ID 0x18.

```
// Turn Signals CAN Protocol
CAN ID = 0x18
Left turn: Byte 5 = 0xC0
Right turn: Byte 5 = 0xA0
```

First, we need to tell ROS to subscribe to CAN messages from socketcan_bridge, and register the message handler as a callback function to process the CAN message. The callback function will be called whenever a new CAN message arrives.

```
ros::Subscriber sub = node.subscribe("sent_messages",
    RECEIVE_BUFFER_SIZE, callback);
```

In the callback function, we can check the value of byte 5 and set the left and right signals accordingly.

```
#include <ros/ros.h>
#include <can_msgs/Frame.h>

void callback(const can_msgs::Frame& msg)
{

    if (msg.id == 0x18)
    {
        // reset turn signals
        left_signal = FALSE;
        right_signal = FALSE;

        // set turn signal according to byte 5 value
        switch (msg.data[5])
        {
            case 0xC0 : left_signal = TRUE;
                        break;
            case 0xA0 : right_signal = TRUE;
                        break;
            default   :
                        break;
        }
    }
}
```

5.2.6 Sending vehicle commands

The OSCC Wiki states that the throttle commands of the Kia Soul are given in the first two bytes of CAN ID 0x62 (see Table 5.1).

```
// Throttle CAN Command
CAN ID = 0x062
Data length: 8 bytes
Transmit Rate: 20 ms
```

So sending the decimal value 32767 (e.g., the binary value 01111111 11111111) will set the pedal throttle to 50 percent, the equivalent of the pedal being pushed half-way down. Before we can send the CAN frame message, we need to register our node to acknowledge that we are publishing an ROS topic.

```
ros::Publisher pub = node.advertise("sent_messages",
    SEND_BUFFER_SIZE);
```

Table 5.1   Table Throttle Command
CAN Protocol

Bit Offset	Length (Bit)	Data
0	16	Pedal[a]
16	8	Reserved
24	1	Enabled
25	1	Clear
26	1	Ignore
27	29	Reserved
56	8	Count

[a] Pedal Command (0 = 0%, 65535 = 100%)
Source: OSCC

Finally, we can fill in the ROS message, publish it, and let the socketcan_bridge node translate the ROS message to CAN frame and send it.

```
#include <ros/ros.h>
#include <can_msgs/Frame.h>

// fill in CAN frame message
can_msgs::Frame msg;
msg.id = 0x062;       // CAN ID of the throttle command
msg.dlc = 8;          // Data length: 8 bytes
// assuming little-endian byte order
msg.data[0] = 0xFF; // binary: "11111111" or decimal: 255
msg.data[1] = 0x7F; // binary: "01111111" or decimal: 127

// send the message
pub.publish(msg);
```

5.2.7   Recording and visualization

Recording the vehicle in operation is a crucial part of vehicle testing, because it allows us to analyze and debug algorithms that need to be modified. Visualization of the vehicle's function can either be done during a test drive (online) or by replaying stored data (offline).

5.2.7.1   Recording and playing back data

To store ROS message data in the ROS middleware, we use the 'bag' file format. Named after its '.bag' extension, this file extension plays an important role in ROS. Various tools are available to store, process, ana-

Figure 5.6: RViz visualization example.

lyze, and visualize this kind of data. Bags are the primary mechanism for data logging in ROS and they have a variety of offline uses.

To record a bag file:

```
$ rosbag record -a -o sample.bag
```

To play back a bag file:

```
$ rosbag play sample.bag
```

5.2.7.2 *Visualization using the RViz tool*

RViz (ROS visualization) is a 3D visualizer that can be used to display sensor data and state information in ROS. Figure 5.6 shows an example of visualization of LiDAR 3D point clouds and image data from the left and right camera of a stereo camera using RViz.

5.3 Testing

Testing is an essential part of the development process, because it allows us to define the quality, reliability, and maintainability of a product. Core SDV technology is realized in the form of software, so software testing typically consumes a large proportion of overall development resources, usually even more than that required for feature implementation. This section provides a brief overview of the software testing activities that are crucial to developing software-driven products such as SDVs.

There are four general levels of software testing: unit testing, integration testing, system testing, and acceptance testing. Let us take a look at each level separately.

5.3.1   Unit testing

The goal of *unit testing* is to ensure that all the components of the software work properly on their own. Each component/module is isolated from the remainder of the code and then tested individually to check that it functions as required. Unit tests are typically run as *white-box* tests, a method that validates the internal structures and workings of the software implementation. To design test cases in white-box testing, you need some knowledge of how the software works or is implemented [2]. Unit testing involves testing a component in isolation; that means you need to mock or simulate all the dependencies (other components on which the test component depends) or replace them with fake components.

If we treat the localization module described in the functional architecture as a single unit or component, then the unit tests for this component might involve the following:

■ Test cases with normal input values, such as all values within the expected minimum and maximum boundaries

■ Test cases with out-of-boundary input values, i.e., values outside the minimum and maximum boundaries

■ Test cases with faulty values, such as sets of values that make no sense semantically (contradiction), or that have a checksum error

5.3.2   Integration testing

The goal of *integration testing* is to test the behavior of all the components when they are integrated or connected to each other. Unlike unit testing, all the components used in an integration test are real software components [2]. It is worth remembering, however, that SDV systems tend to be very complex, with a lot of dependent components, so it may be necessary to divide the system into subsystems to make it easier to analyze faults when unexpected behavior occurs. Running an integration test on a subsystem means that all the components under test in the subsystem are real components, while the rest may be simulated.

Using the localization module described above as an example, one subsystem might be the sensor fusion, which comprises a GNSS-based positioning component, an IMU and odometry-based positioning component, a camera-based positioning component, and a lidar-based positioning component. Depending on which middleware and technology you are using, the interaction between the components might be realized via message passing, shared memory variables, function calls, or other mechanisms of data interchange. The behavior of each component is defined in an interface that specifies the valid range of input and output values, expected input and output data structure, cycle time (if the component runs periodically), and so on. Examples of integration testing include the following:

- Test cases with valid interface values

- Test cases with faulty interface values, i.e., values outside the valid value range or with a checksum error

- Test cases with timing errors, such as no input values provided within the expected time window (timeout)

Unit and integration testing focus solely on the software, which is why they are sometimes referred to as *Software-in-the-Loop (SIL)* tests. Generally speaking, SIL tests do not require special hardware because the test environment and all the test cases are purely software. That means you can run most SIL tests on a personal computer (PC) using a standard desktop operating system (OS) such as Windows, Linux or macOS. This can be the same PC used to develop the software, though it is good practice to run the SIL test environment on a dedicated server, so that it can be easily accessed by the whole development team, and executed regularly, and automatically.

5.3.3 System testing

The next level is *system testing*, which involves testing the functionality of the whole SDV software system in conjunction with all the other systems in the vehicle, such as the vehicle gateway, and the vehicle platform. The aim of system testing is to verify the correct behavior of the entire vehicle before testing it on the road. This level assumes no knowledge of implementation detail. The system is regarded as a black box that communicates with other systems on the vehicle bus using real bus messages or signals (CAN, Ethernet, etc.). System testing may include the following:

- Test cases with simulated driving situations, such as obstacle avoidance tests with simulated sensor values representing pedestrians or other vehicles

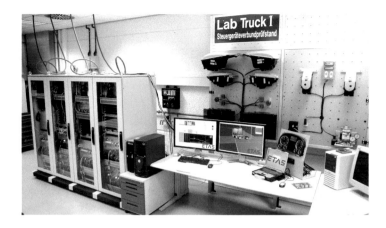

Figure 5.7: HIL test environment using a cluster of ECUs. (Reprinted with permission from ETAS GmbH. ©2009 ETAS GmbH)

- Test cases with simulated faulty components or communication, i.e., test cases with blocked or temporarily unavailable sensors, tampered bus messages, or signal timeouts

- Stress tests, such as test cases that generate maximum processor/memory bus load or run for extended periods

This phase involves testing the SDV software on its physical computing platform hardware in conjunction with other hardware in the vehicle, so we typically use a *Hardware-in-the-Loop (HIL)* test environment. Unlike SIL, HIL tests require at least one PC or server (to control test execution) as well as a separate piece of hardware, either the actual computing platform hardware on which the SDV software runs, or some other form of hardware that emulates the computing platform. Whereas a *Closed-Loop-HIL* test environment also uses some or all output data of the *System Under Test* as part of its input, no output of an *Open-Loop-HIL* test setup is fed back into the input data.

Sometimes, it is impractical or simply too expensive to rig up an HIL test environment with all the vehicle hardware. As an alternative, the HIL test environment could comprise a small subset of the vehicle hardware, as depicted in Figure 5.7, or even just the computing platform hardware itself, while the rest of the hardware could be simulated using Rest Bus Simulation (RBS) software. Depending on the complexity of the test cases and the availability of testing resources, test execution can also be automated.

5.3.4 Acceptance testing

The final step of the software testing process is conducted directly on the vehicle itself. The main focus here is to test the SDV product as a whole and ensure that the overall product meets users' or customers' expectations. The *acceptance test* is a useful way of identifying technical issues caused by incomplete or faulty specifications, incorrect implementations that were not detected at other test levels, and areas for improvement. As well as random testing, such as test drives along random routes or in random situations, acceptance test cases normally include test drives based on pre-defined scenarios on pre-defined ('reference') routes, enabling the results to be compared with previous data. Long-term system stability and behavior correctness over a long period of driving in real traffic situations are also important goals for this test level. Acceptance tests are usually carried out manually because they typically involve human (subjective) feedback as shown in Figure 5.8. Other test cases at this level might also include:

- Test drives in various weather conditions, such as snow or rain

- Test drives in different geographical regions, i.e., North America, Western Europe, Southeast Asia, etc.

- Test drives in extreme climate/temperature conditions, i.e., summer testing in a desert in the Middle East, or winter testing in Scandinavia

Figure 5.8: Acceptance test performed in the vehicle. (Reprinted with permission from Continental AG. ©2017 Continental AG)

5.4   Summary

This chapter explored some of the issues you need to consider when you start to construct an SDV prototype. Your first step is to choose a suitable vehicle. This needs to either already provide by-wire control, or you will need to retrofit this to the vehicle. The next step is to ensure you can access the vehicle's network in order to get access to the in-built sensors. This may require you to analyze the bus messages in order to understand what the data means.

Once you have a vehicle that you can control, and whose sensors you can access, the next stage is to choose the exteroceptive sensors that you will add. Key considerations here are the operating environment, the operating conditions, and the budget. Having installed the sensors, you will need to calibrate them, and test that they are working correctly. The next stage of the development process is to select a computing platform and install the required middleware and control software. In Section 5.2.2 we showed examples of how to do this for the ROS middleware.

We looked in detail at the OSCC, which provides the required hardware, firmware and software to allow you to convert a certain car into an SDV prototype. This approach can save a huge amount of time as well as being much cheaper than developing a prototype by yourself. We then looked at how to install the required middleware and drivers on the computing platform. In Section 5.2.3, we explained the different approaches for development. Hand-coded development involves a software developer writing the code from scratch, while model-based development uses tools such as MATLAB Simulink to generate the code from a model of the system. Implementing all the required algorithms, such as multi-sensor data fusion and vehicle control, can be a challenging task, and often needs a combination of the two approaches.

Once you have developed your software, you need to test it. Section 5.3 explains the different levels of testing and shows some sample test cases for each level. Software in the loop testing involves progressively testing individual units/functions, then integrating these into larger functions. Hardware in the loop testing then takes these and tests them on the actual hardware (potentially simulating things like sensors), before finally you are able to test that the complete vehicle performs as expected and is acceptable to the 'customer'.

In the next chapter we will be looking at some of the other issues that have to be considered when creating an SDV. Chief among these are safety (clearly, you do not want to create an SDV that is unsafe), security (ensuring your SDV cannot be hijacked by a malicious 3rd party), and the need for backend systems, such as a source of mapping data, traffic data, etc.

References

[1] Jonny Andersson. Entwicklung eines notbremssystems bei scania. *ATZelektronik*, 12(1):36–41, Feb 2017.

[2] Kai Borgeest. *Software*, pages 213–277. Springer Fachmedien Wiesbaden, Wiesbaden, 2008.

[3] Manfred Broy, Sascha Kirstan, Helmut Krcmar, and Bernhard Schätz. What is the benefit of a model-based design of embedded software systems in the car industry? In *Software Design and Development: Concepts, Methodologies, Tools, and Applications*, pages 310–334. IGI Global, 2014.

[4] J. M. Collado, C. Hilario, A. de la Escalera, and J. M. Armingol. Self-calibration of an on-board stereo-vision system for driver assistance systems. In *2006 IEEE Intelligent Vehicles Symposium*, pages 156–162, June 2006.

[5] Jamar Gibson. State laws and regulations and local initiatives. `https://www.johndaylegal.com/state-laws-and-regulations.html`. [Online; accessed 08-Jan-2018].

[6] HERE. Here HD live map. `https://here.com/en/products-services/products/here-hd-live-map`. [accessed 08-Jan-2018].

[7] Jackie Kay and Adolfo Rodriguez Tsouroukdissian. Real-time control in ros and ros 2 - roscon 2018. `https://roscon.ros.org/2015/presentations/RealtimeROS2.pdf`. [accessed 08-Jan-2018].

[8] MISRA. Guidelines for the use of the C language in critical systems. *MIRA Limited. Warwickshire, UK*, 2004.

[9] ROS-Wiki. How do I build ros vlp16 velodyne driver for indigo using catkin edit. http://answers.ros.org/question/226594/. [accessed 08-Jan-2018].

[10] Tomtom. Tomtom hd map roaddna | tomtom automotive. https://www.tomtom.com/automotive/automotive-solutions/ automated-driving/hd-map-roaddna/, 2018. [accessed 08-Jan-2018].

[11] H. Wei, Z. Huang, Q. Yu, M. Liu, Y. Guan, and J. Tan. Rgmp-ros: A real-time ros architecture of hybrid rtos and gpos on multi-core processor. In *2014 IEEE International Conference on Robotics and Automation (ICRA)*, pages 2482–2487, May 2014.

[12] Mike Whalen. Why we model: Using mbd effectively in critical domains. *Workshop on Modeling in Software Engineering @ ICSE 2013*, 2013.

Chapter 6

Other technology aspects

Building an SDV is not simply a matter of solving the technological issues already discussed. If you want to go from simple prototype to a production vehicle, there are other, often external, issues that also have to be taken into account. Chief among these is the need to ensure the safety of the vehicle, from both the functional viewpoint and the cybersecurity viewpoint. SDVs can only hope to win public acceptance if they are at least as safe as conventional vehicles, and thus it will be essential to protect them from external attacks of the sort that have bedeviled users of the Internet over recent years. This is especially important, given the fact that most SDVs will need to rely on external data sources, such as back-end systems, as well as V2X (Vehicle-to-Everything) networks which will provide up-to-date information on their immediate driving environment. In this chapter we will look at these issues in more detail.

6.1 Functional safety

Functional safety is defined in *ISO 26262* as 'the absence of unreasonable risk due to hazards caused by malfunctioning behavior of electronic and/or electric (E/E) systems' [24]. Hazards are anything with the potential to cause people physical injury or to damage their health. The more driving tasks that are automated, the higher will be the potential safety risks due to system malfunction. Thus, functional safety is one of the most important aspects throughout the entire SDV development lifecycle.

6.1.1 Why is functional safety important?

Functional safety is a crucial aspect of any application that involves E/E systems. In vehicles it is particularly important because system malfunction may lead to humans both inside and outside the vehicle being injured or even killed. Functional safety compliance gives customers the assurance that the vehicles are safe to use, and meet a commonly accepted safety standard, regardless of the vehicle brand, model, and technology details. This is especially important for SDVs, where the success of the technology relies on it performing at least as well as existing technology, and its public acceptance correlates directly with how safe society perceives that technology to be.

Functional safety standards are not only important for consumers, but also for manufacturers. The standards establish the state-of-the-art safety processes, requirements, and guidelines, based on current best practices. Failing to meet the current standards leaves manufacturers open to product liability claims. In Germany, for example, the first article of the product liability law states that, *'... The compensation obligation of the manufacturer is only excluded, ... if the fault could not be known according to the state of science and technology at the time the manufacturer put the product into distribution.'* [6]. Complying with industry-wide functional safety standards such as *ISO 26262 (Road Vehicles – Functional Safety* and *IATF 16949 Quality Management for the automotive industry)* help manufacturers minimize risks of product liability. However, compliance with functional safety standards does not automatically exempt vehicle manufacturers from future liability issues. In the U.S., consumer protection laws vary from state to state. Asserting 'state-of-the-art' might be a valid defense in product liability case in some states, but not in others [47]. As technology evolves, standards might become obsolete as they no longer reflect the current state of technology. Therefore, functional safety standards, such as ISO 26262, should be regarded as minimum safety requirements [41].

6.1.2 ISO 26262

The ISO 26262 Road Vehicles—Functional Safety standard is the automotive adaptation of the International Electrotechnical Commission IEC 61508 general industry standard for functional safety [19]. In its ten parts, it outlines functional safety management, engineering processes, recommendations for different phases in product development using the *V-model* as reference process model, and the supporting processes. V-model is a standard process model in automotive system engineering that represents the development activities using a V shape, with specification and design on the left side, test and integration on the right side, and implementation at the bottom of the V. The V-model shows the direct relationship between each phase of the development activities and its corresponding phase of testing. Figures 6.1 and 6.2 show an overview of the ISO 26262 standard and V-model, respectively.

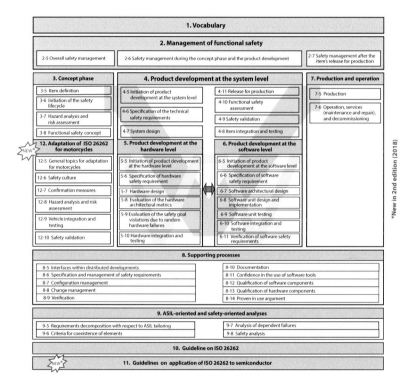

Figure 6.1: ISO 26262 parts overview. The second (2018) edition of the standard includes two new additional parts, namely "Guidelines on application of ISO 26262 to semiconductors" and "Adaption of ISO 26262 for motorcycles." (Adapted from "Road vehicles - Functional safety - Part 1: Vocabulary", by International Standard Organization (ISO), 2011, ISO 26262-1:2011, p. vi. ©2011 ISO)

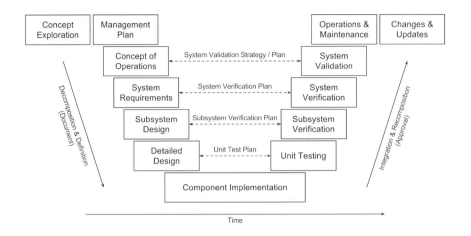

Figure 6.2: V-model. (©Behnam Esfahbod, https://commons.wikimedia.org/ wiki/File:Vee_Model_for_Systems_Engineering_Process.svg, "Vee Model for Systems Engineering Process", https://creativecommons.org/licenses/by-sa/ 3.0/legalcode)

6.1.2.1 Safety management

The first two parts of the ISO 26262 standard cover vocabulary and safety management. Safety management is about performing safety activities throughout the 6 phases of the safety lifecycle. These are concept, development, production, operation, service and decommissioning phases. Tailoring of the safety activities is allowed by the standard, as long as a rationale for non-compliance or non-application exists, and it has been assessed according to the standard. The recommendations and requirements for the safety management are divided into three categories, namely the overall safety management, safety management during concept and development phase, and safety management from production onwards. The overall safety management requirements include the assessment of the organization's safety culture, competence management (to ensure persons involved have sufficient level of skills, competences and qualifications), and compliance with a common quality management standard, such as *IATF 16949* or *ISO 9001*.

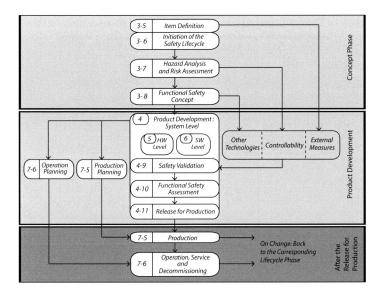

Figure 6.3: Reference safety lifecyle. (Adapted from "Road vehicles - Functional safety - Part 2: Management of functional safety", by International Standard Organization (ISO), 2011, ISO 26262-2, p. 4. ©2011 ISO)

6.1.2.2 Engineering processes and requirements

The next parts of the standard cover the engineering process and requirements throughout the safety lifecycle as shown in Figure 6.3, starting with the concept phase. The safety activities in the concept phase involve the definition of the *item*, i.e., the system to be considered at vehicle level (airbag, electronic brake, etc.), and the initiation of the safety lifecycle, i.e., distinguishing between new item development or modification of an existing item. In case of modification, an impact analysis that includes the identification of the intended modification, assessment of the impact of the modification, and optionally, the tailoring of the safety activities are part of the refined safety plan. The aforementioned safety activities are followed by *Hazard Analysis and Risk Assessment (HARA)*. HARA is a method to systematically identify and classify the *hazardous events*, i.e., scenarios where hazards can occur during a vehicle's life. This is done with the help of common techniques, such as *Failure Mode and Effects Analysis (FMEA)* or *Failure Tree Analysis (FTA)*. The aim is to determine the *Automotive Safety Integrity Level (ASIL)*, and the safety goals to mitigate the associated hazards. Finally, the concept phase concludes with the

Table 6.1   ASIL determination table

Severity (S)	Exposure (E)	Controllability (C)		
		C1	C2	C3
S1	E1	QM	QM	QM
	E2	QM	QM	QM
	E3	QM	QM	A
	E4	QM	A	B
S2	E1	QM	QM	QM
	E2	QM	QM	A
	E3	QM	A	B
	E4	A	B	C
S3	E1	QM	QM	A
	E2	QM	A	B
	E3	A	B	C
	E4	B	C	D

Source: Adapted from "Road vehicles - Functional safety - Part 3: Concept phase", by International Standard Organization (ISO), 2011, ISO 26262-3, p. 10. ©2011 ISO.

definition of the functional safety requirements derived from the safety goals.

6.1.2.3   *Automotive safety integrity level*

Automotive Safety Integrity Level (ASIL) is a standard way to assess the necessary safety measures that need to be applied to an item in order to avoid unreasonable residual risks. ISO 26262 defines four ASIL levels, ranging from ASIL A (the least stringent level) to ASIL D (the most stringent level). ASIL D functions require more comprehensive safety requirements and measures than ASIL A, B and C. Many functions in the vehicle, such as entertainment applications under the domain of Information and Entertainment (Infotainment) Systems, are not safety relevant, and therefore do not require safety measures as defined in ISO 26262. These functions are classified as ASIL QM, as only standard Quality Management (QM) processes are applied to them.

The ASIL level is determined from the *Exposure (E)*, *Controllability (C)*, and *Severity (S)* of each hazardous event according to the ASIL determination table. Exposure refers to the probability of the hazardous event occuring. Controllability is associated with the ability to avoid the harm related to the event. Severity quantifies the seriousness of the consequences caused by the situation. The intersection of the severity, probability (exposure), and controllability defines the ASIL level as shown in Table 6.1.

6.1.2.4 Product development

The concept phase is followed by the product development phase. At the system level, the functional safety concept is now refined into technical safety requirements specifications that define the necessary safety mechanisms to achieve the associated functional safety requirements. For example, the specification defines the safety measures to detect and control faults in the system itself, or in external systems, and the safety measures to achieve and maintain a *safe state*, i.e., an operating state that does not have unreasonable level of risk. The next step is to develop the system design, and the technical safety concept based on these technical safety requirements specifications. The technical safety concept contains the safety measures for avoiding system failures, and controlling random hardware failures during vehicle operation. Part of the technical safety concept relates to the allocation of technical safety requirements to hardware and/or software, the specification of a Hardware-Software-Interface (HSI), as well as the system level requirements for production, operation, service, and decommissioning.

During the development phase, both the hardware and software need safety requirements, architectural designs, detailed/unit designs and safety analysis (to identify possible causes of failures and their effects), integration and integration tests. Finally, safety activities at system level conclude with the integration and the validation phases. The integration phase begins with planning of system integration, and planning of integration verification, integration of hardware and software, system integration, and (system) integration tests, as well as vehicle integration, and vehicle testing. The validation phase provides requirements related to validation planning, as well as release documentation.

6.1.2.5 Production and the safety lifecycle

After development, the standard specifies requirements and recommendations, on phases after the item's release for production. The standard outlines three sub-phases of the production phase: planning, preproduction and production. The recommendations and requirements in the production phases are designed to meet two objectives: to maintain a production process for safety-related elements/items to be installed in vehicles; and to achieve functional safety during the production process. Finally, during the operation, service, and decommissioning phases, the standard provides requirements and recommendations for maintenance planning and repair instructions, the warning and degradation concept, field monitoring processes, decommissioning instructions, and other requirements to maintain functional safety throughout the safety lifecyle.

6.1.2.6 Supporting processes

The last parts of the standard cover requirements for the supporting processes, ASIL-oriented and safety-oriented analysis and guidelines on applying the standard. Supporting processes refer to processes/activities that are not specific to one particular safety lifecycle phase, but are essential for achieving functional safety, and traceability. The supporting processes include the correct specification and management of safety requirements, configuration management, change management, qualification of hardware/software components, etc. The ASIL-oriented and safety-oriented analyses part covers the topic of *ASIL decomposition*, coexistence of sub-elements that have different ASIL levels assigned, analysis of dependent failures and safety analysis. The final part of the standard has an informative character that aims to enhance understanding by providing a general overview of ISO 26262, as well as examples and additional explanations on selected parts of the standard.

ASIL decomposition is an ASIL tailoring technique to achieve a given safety goal using several independent elements that address the same goal, but with possibly lower ASILs. ASIL decomposition uses a notation 'ASIL C(D)', which means an ASIL C requirement that is part of a target ASIL D decomposition. For example, an ASIL D requirement can be addressed combining an ASIL C(D) requirement with an ASIL A(D) requirement, or two ASIL B(D) requirements. The table of permitted combinations is provided in the penultimate part of the ISO 26262 standard [23]. ASIL decomposition is useful in situations where the effort or development cost of achieving the two decomposed functional safety requirements is lower than that of achieving the original one.

6.1.3 Challenges

Complying with ISO 26262 takes significant effort and resources throughout the automotive safety lifecycle, and applying the standard to SDV functions is no exception.

According to Spanfelner et al. [42], one of the big challenges lies within the general issue of 'insufficient models'. In order to derive the complete safety measures required to achieve functional safety, a complete system model of the function is needed, which includes all external factors and their influences on the function. However, it is simply not possible for some SDV functions to have such a complete model. Take a function that tries to predict the behavior of pedestrians, for example. It is impossible to list all external factors, and define how they actually affect the function, because they may never be understood. Thus, such models are most likely probabilistic models, which suffer both oversimplification (the reality is far more complex), and underspecification (not all external factors can be identified).

Another challenge is the lack of specification of standard tolerable failure rates for systems that are based on probabilistic models, e.g., object classification, because the failure rates mentioned in ISO 26262 are applicable only to random hardware failures [45].

Furthermore, the emergence of machine learning based functions, such as deep learning and the end-to-end learning approach (which will be discussed briefly in Chapter 7), poses a new challenge to functional safety. In such systems, the whole decision-making process is done by the computer based on an internal system model, which is computer generated and not understandable (or even visible) to humans. Thus, a new approach or paradigm for functional safety might well be necessary.

6.2 Cybersecurity

The terms security and safety are sometimes used interchangeably, even though they technically have different meanings. Some languages even use the same word for both security and safety, such as the German word 'Sicherheit'. In the context of electronic/electrical systems such as SDVs, safety usually means the absence of unreasonable risks due to system malfunction, and thus implies protection against unintentional events. Security, on the other hand, is associated with protecting the system from the intentional exploitation of vulnerabilities through cyber attacks. Another aspect that is sometimes used to highlight the difference between the two is the source of threats. Safety issues mostly arise from within the vehicle, whereas security issues are mostly caused by external factors or from events outside the vehicle.

6.2.1 Why is cybersecurity important?

In order to understand the importance of cybersecurity, it might be helpful to see security in relation to safety. 'A safety-critical system is a security-critical system, but not all security-critical systems are safety-critical'. The US-based Society of Automotive Engineers (SAE) states this strong relationship between safety and security in their *Cybersecurity*

Guidebook for Cyber-Physical Vehicle Systems (SAE J3061) [37] (see Figure 6.4). In other words, functional safety alone is not enough to make vehicles safe, if it is not complemented with security by design.

Figure 6.4: Safety-critical and security-critical systems relationship. (Adapted from "Cybersecurity Guidebook for Cyber-Physical Vehicle Systems", by Society of Automotive Engineers (SAE) International, 2016, J3061, p. 17. ©2011 ISO)

While conventional, or non self-driving, vehicles may still function well without requiring any external connectivity, this is unlikely to be the case for an SDV. In fact, the higher automation level an SDV has, the more likely the vehicle needs to rely on external information, e.g., from back-end systems or other sources. This communication with the 'outside world' also exposes SDVs to potential cyber attacks, ranging from an attempt to compromise privacy, e.g., stealing personal data, to vehicle manipulation, e.g., deactivating some driving functions that may cause harm to people inside the vehicle and in its surrounding environment [25].

6.2.2   *Automotive cybersecurity standards*

Computer security standards have been evolving since the 1950s, pioneered by a standard initiated by the US government to limit the allowable emanation levels of mainframe radiation to protect computer systems from eavesdropping attacks [36]. However, cybersecurity standards/guidelines for the automotive industry only began emerging around the beginning of this millennium.

One of the early automotive cybersecurity standardization initiatives was the Secure Vehicle Communication (SeVeCom) project. The main objective of this EU-funded project was to define a reference architecture for secure Vehicle-to-Vechicle (V2V) and Vehicle-to-Infrastructure (V2I) networks. Furthermore, the project also identified priority and longer-term research areas to enable secure V2X networks as well as the deployment roadmap for security functions on these networks [27].

Another noteworthy European project with regard to automotive cybersecurity was the E-Safety Vehicle Intrusion Protected Applications (EVITA) project. In contrast to the SeVeCom project, the EVITA project focused on designing, verifying and prototyping a secure automotive on-board network architecture [18]. The EVITA Hardware Security Module (HSM) specification, also known as the EVITA HSM standard, has become one of the major HSM standards in the automotive industry. EVITA HSM will be covered in more detail later.

SAE J3061 [37], mentioned in the last section, is the first cybersecurity engineering guideline for the automotive industry. First published in early 2016, SAE J3061 provides best practices for the development of security-critical automotive applications. The guideline is based on ISO 26262 and uses a similar development process for security engineering in all phases of vehicle lifecycle. SAE J3061 serves as basis for the new international automotive cybersecurity standard ISO/SAE 21434 Road Vehicles-Cybersecurity Engineering, currently (at the time of writing) under joint development by the SAE and ISO standards organizations [22].

Besides other works initiated by automotive consortiums and standardization bodies, new approaches have also been proposed by combining well-known methodologies from the 'classical' IT world, and best practices for automotive systems engineering. One such example is the *Security-Aware Hazard and Risk Analysis (SAHARA)* approach, that combines the popular *STRIDE* threat modeling approach from the IT world with the widely known *HARA* methodology for functional safety [29]. STRIDE stands for Spoofing, Tampering, Repudiation, Information disclosure, Denial of Service and Elevation of privilege. It is also known as the 'Microsoft threat model', a widely used threat modeling method developed by Microsoft employees Loren Kohnfehlder and Praerit Garg in the late 1990s [26]. HARA, or Hazard Analysis and Risk Assessment, is a standardized safety analysis method, specified in part 3 of ISO 26262 specification, used for determining safety goals and provides the basis for deriving functional safety requirements from the safety goals [24].

6.2.3 Secure SDV design

In the following section, we will investigate some of the system design considerations for developing secure SDVs using a multi-leveled system approach, i.e., hardware, software, in-vehicle network communication and external communication levels.

6.2.3.1 Secure hardware

This level of security focuses on the protection of physical vehicle components from external manipulation or unauthorized access. Hardware level security is typically enforced with the support of a *Hardware Security Module (HSM)*. HSMs comprise a *cryptographic service engine* (typically hardware-accelerated) and a *secure key storage*. Cryptographic functions, such as data encryption/decryption, and message digest calculation, are naturally resource-intensive computations, and therefore are better moved to a dedicated component. The secure key storage protects security keys from illegal access or tampering. HSM may also support *secure boot*, a mechanism to avoid a tampered boot loader from running by verifying the digital signature of the code prior to booting.

There are several major hardware security standards widely used in the automotive industry, in particular the EVITA HSM, SHE and TPM standards.

The *E-Safety Vehicle Intrusion Protected Applications Hardware Security Module (EVITA HSM)* standard, as already mentioned, was published by the EVITA consortium in 2011. The EVITA HSM standard specifies three HSM profiles (or versions): *light*, *medium* and *full*. The light EVITA HSM profile specifies only an internal clock, basic hardware-accelerated cryptographic processing, usually the symmetric encryption/decryption algorithm according to the *Advanced Encryption Standard* with 128 bit key (AES-128), as well as a physical *True Random Number Generator (TRNG)*, that can be used by the built-in *Pseudo-random Number Generation (PRNG)* algorithm. The light profile is designed to enable secure communication in components with cost and efficiency constraints, such as in sensors and actuators [46]. The medium profile is intended to enable secure in-vehicle network communication networks and adds several additional requirements on top of the light profile, such as secure ticks (monotonic counters), secure memory, secure boot mechanism, as well as support for cryptographic hash functions, e.g., the *Secure Hash Algorithm (SHA)*. Finally, the full profile provides support for highly demanding automotive cybersecurity applications, such as secure and time-critical communication within V2X networks. At this level, the cryptographic function

Table 6.2   EVITA HSM profiles comparison

Aspect	Full	Medium	Low
Internal RAM[a]	✓	✓	optional
Internal NVM[b]	✓	✓	optional
Symmetric Cryptographic Engine	✓	✓	✓
Asymmetric Cryptographic Engine	✓	-	-
Hash Engine	✓	-	-
True Random Number Generator	✓	✓	optional

[a] *Random Access Memory*
[b] *Non-Volatile Memory*

Source: Adapted from "Secure Automotive On-board Electronics Network Architecture", by Ludovic Apvrille, Rachid El Khayari, Olaf Henniger, Yves Roudier, Hendrik Schweppe, Hervé Seudie, Benjamin Weyl, Marko Wolf, 2010, p. 5.

is performed by a high-performance 256-bit asymmetric cryptographic engine based on high-speed elliptic curve arithmetic [17]. It also replaces the hash function with an AES-based function called *WHIRLPOOL* [34]. Table 6.2 presents some key differences between the EVITA HSM profiles.

The *Secure Hardware Extension (SHE)* was proposed by the German car manufacturers consortium HIS (Hersteller-Initiative Software) in 2009. SHE is designed as a low-cost secure key storage and cryptographic service engine, which is typically implemented as an on-chip extension to existing electronic control units (ECUs) [39]. With regards to functionality, SHE is quite similar to the EVITA HSM light specification. However, unlike EVITA light, SHE provides secure boot as standard.

Another major HSM standard is the *Trusted Platform Module (TPM)* developed by the Trusted Computing Group (TCG). TPM is standardized as ISO/IEC 11889 and is widely known due to the ubiquitous use of TPM chips in modern PCs, and laptops. The *TPM 2.0 Automotive-Thin Profile*, initially released in 2015, specifies a subset of the TPM 2.0 specification that is suitable for deployment in resource-constrained ECUs [16]. Similar to the EVITA HSM and SHE standards, TPM Automotive-Thin Profile also supports secure key storage and management. However, with TPM, the emphasis is more on protecting firmware and software integrity, supporting software attestation and enabling secure software updates, rather than providing hardware-based support for secure in-vehicle network communication.

6.2.3.2 Secure software

An SDV is a highly complex system, whose usability is largely determined by the quality of its software. As we have seen throughout this book, an SDV's software is a collection of many specialized functions that each solve a particular task. Each of these functions is built upon other software, and in turn each of its dependencies relies on other software, and so on. The sheer number of dependent software components integrated from various sources in an SDV seems unavoidable, but it also creates security risks. A chain is only as strong as its weakest link, and the same principle also applies to security: the whole system will only be as secure as its most vulnerable component. However, keeping all software components secure at all times is hardly an easy task, due to the high complexity of the system. Best practices in software development can help minimize security risks in own-developed software. These include defensive programming, peer code review, along with the integration of static code analysis, data flow analysis, code complexity analysis, and other tools as an integral part of the software development workflow. However, this might prove difficult, if not impossible, to enforce on proprietary third-party software. Due to the lack of source code or any detailed information on the software, and its dependencies, security patches to the component and its dependencies can only be made available by the software vendors, since only they know the implementation details, and any dependencies that need to be updated.

Creating a secure SDV is not only the result of developing the system with security in mind, and keeping it secure throughout the whole vehicle product lifecycle. It is also an outcome of many careful and thorough security-relevant decisions during the design phase. Similar to functional safety, cybersecurity should never be thought as an 'add-on' to be put off until a later development stage, because poor decisions made early on in the design stage might be very costly to reverse later in development. In some cases they might even risk the failure of the project, product usability or acceptance as a whole.

One example of such early considerations is the choice of hardware platform and operating system. Common software security features, such as secure boot and secure debug, require hardware that support these functions. *Secure debug* is a means to debug (locate errors in) software in an ECU securely at run-time. *Software partition* is a general technique for splitting software parts or functionalities into several isolated instances to minimize risk of interference. A closely related security method, called *embedded virtualization*, uses an embedded hypervisor to efficiently run multiple isolated virtual machines on a single embedded system. Figure 6.5 shows an example of embedded virtualization use case, where a security breach in one VM instance does not compromise the security of other VM instances. However, in order to take the full advantage of embedded virtualization requires a combination of suitable hardware, e.g., one that has at least a Memory Protection Unit (MPU), and a suitable Real-Time/Embedded Operating System (RTOS) that supports or acts as a secure and efficient embedded hypervisor.

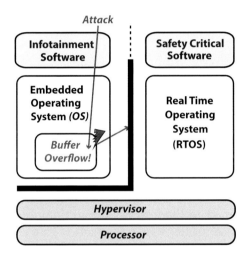

Figure 6.5: Embedded virtualization use case. A security breach in one virtual machine does not compromise the security of other virtual machine instances. (Adapted from "The role of virtualization in embedded systems", by Gernot Heiser, 2008, doi: 10.1145/1435458.1435461, 04, p. 11-16)

Last, but not least, another critical security area in SDV software is related to the detection of sensor attacks, and minimizing the impact of such attacks. An SDV relies heavily on the information obtained from its sensors, so effective countermeasures to this kind of attack could be life saving. Sensor attacks can come in many forms. A *spoofing attack* involves generating fake sensor signals, which lead the target sensor to believe the presence of something that actually does not exist in reality. The aim of a *jamming attack* is to distort input signals from sensors, so that the real signals cannot be reliably distinguished from noise anymore. A *blinding attack* works by shining an intense light directly at a camera to impair its visibility, or even to damage the sensor permanently. Another form of attack is by capturing pulses transmitted by the target sensor, and resending them at different times (*replay attack*) or resending them from different positions (*relay attack*). Successful sensor attacks on radar, ultrasonic sensors and cameras have been reported in [48]. Some feasible and effective attacks on cameras and lidar, as well as proposed countermeasures, are shown in [33]. GNSS receivers have long been known to be vulnerable to attacks. An overview of GNSS threat scenarios and an evaluation of known defense mechanisms are presented in [40].

6.2.3.3 Secure in-vehicle network communication

In 2015, *WIRED* magazine published a far-reaching security article describing how two security experts, Miller and Valasek, gave a live demonstration in which they successfully took over a running vehicle from a remote distance, and left the driver with no chance to counter the attack [15]. Later that year, the two experts published a document that shows the technical details of how their remote attack was made possible. In a nutshell, the attack exploited a security hole in the in-vehicle infotainment system that enabled them to trigger an unauthenticated firmware update of the microcontroller which interacts with the vehicle CAN bus. Using the modified firmware, fake CAN messages can then be 'injected' into the vehicle bus in order to override the actual vehicle control [31].

The incident, also known as the 'Jeep hack attack', shows how life threatening an insecure vehicle can be, and how essential secure communication between ECUs throughout the vehicle's network is. Undoubtedly, securing the in-vehicle network alone won't be enough to prevent such attack. Other vulnerabilities, e.g., unauthenticated software updates, insecure WiFi password generation mechanisms, open diagnostic ports, etc. need to be addressed as well. Having said that, we will focus on securing in-vehicle network communication, using an approach recommended by AUTOSAR (the Automotive Open System Architecture, see Section 4.3.3).

AUTOSAR security modules

There are two modules in AUTOSAR that are responsible for securing communication throughout the vehicle network. The *Secure Onboard*

Communication (SecOC) module is responsible for generating and veri-fying secure messages passed between ECUs over the in-vehicle com-munication network. The AUTOSAR SecOC specification is designed with 'resource-efficient and practicable authentication mechanisms' in mind, so that legacy systems can also profit from it with minimal overhead [3]. The *Crypto Service Manager (CSM)* module provides basic cryptographic services, e.g., encryption/decryption, Message Authenti-cation Code (MAC) generation/validation, etc. to all modules at run-time, including the SecOC module [4]. Depending on the platform used, some cryptographic functions might use hardware implementations, e.g., HSM/SHE, or software implementations as part of the AUTOSAR Basic Software stack. In this case, the CSM provides an abstraction layer to all cryptographic functions, so that all AUTOSAR modules can use the same standardized API regardless of the implementation detail.

SecOC overview

Within the AUTOSAR stack, the SecOC module resides on the same level as the AUTOSAR *Payload Data Unit Router (PduR)* module within the Basic Software Layer (see Figure 6.6). Payload Data Unit (PDU) is a gen-eral term for data exchanged across the vehicle network. In a nutshell, the PduR module distributes PDUs transmitted over various vehicle buses, e.g., CAN, FlexRay, Ethernet, etc. to AUTOSAR modules, and vice versa, based on a statically configured routing table. On the incoming of secure PDUs, i.e., PDUs whose communication needs to be secured since not all PDUs are security-relevant, the PduR pass the message to be verified by the SecOC module and only if the verification is successful, the PDU will be routed to the higher AUTOSAR modules as usual. And before a secure PDU is transmitted to the vehicle bus, the SecOC module adds some secu-rity data to the PDU for authentication purpose on the receiver data. It is worth noting that the whole security mechanism is performed at a quite low level in the AUTOSAR communication stack, and is therefore com-pletely transparent to the user applications, i.e., SWCs on the application layer. In other words, no modification to user applications is necessary to take advantage of the secure in-vehicle network communication.

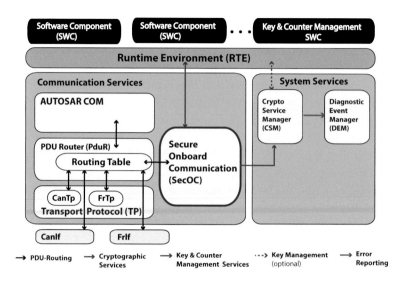

Figure 6.6: SecOC module within the AUTOSAR Basic Software stack. (Adapted from "Specification of Secure Onboard Communication AUTOSAR CP Release 4.3.1", by Automotive Open System Architecture (AUTOSAR), 2017, p. 7)

In order to prevent spoofing, tampering and replay attacks, the SecOC specification recommends authentication using a *Message Authentication Code (MAC)*, in conjunction with a monotonic counter, also known as *Freshness Value (FV)*. In practice, the *Cipher-based MAC (CMAC)* based on the *Advanced Encryption Standard (AES)* using a 128-bit secret key (AES-128) as specified in the NIST SP 800-38B specification [9] is commonly used. The secret key is unique for every vehicle, and is typically generated using a combination of the unique Vehicle Identification Number (VIN) with some random number. The vehicle-specific secret key makes it difficult to expand attacks to other vehicles (including those of the same model), even if the security of one particular vehicle has been compromised.

As shown in Figure 6.7, the first step of CMAC calculation is to derive sub-keys from the symmetric key using the sub-key generation process. The message is partitioned into a sequence of equal-sized data blocks, and AES symmetric encryption is applied sequentially to each data block, each being XOR-ed with the previous AES result. The MAC is obtained by truncating the last encryption result according to the MAC length parameter. The MAC and the FV are transmitted along with the actual data payload. In practice, these values are not transmitted in full-length, but are truncated. The actual length of the truncated MAC and FV is dependent on the available data length with regard to the protocol being used (for example, CAN-FD supports at maximum 64 bytes in a single message), and is a classical trade-off between security and efficiency.

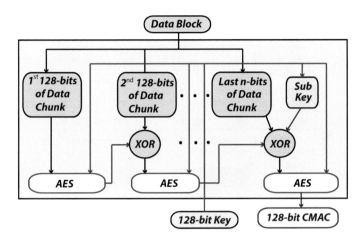

Figure 6.7: CMAC algorithm. (Adapted from "Survey and comparison of message authentication solutions on wireless sensor networks", by Marcos Simplicio, Bruno Oliveira, Cintia Margi, Paulo Barreto, Tereza Carvalho, Mats Naslund, 2013, doi: 10.1016/j.adhoc.2012.08.011, 11, p. 1221-1236)

To further improve security, the SecOC specification recommends MAC calculation using PDU-specific keys. Each PDU (protocol data unit, or bus message) is assigned a unique Data Identifier (Data-ID). Using the Data-ID and the secret key, the individual AES secret key can be derived for the CMAC calculation for that particular secured PDU. So even if the car-specific key has been compromised, it is still difficult for attackers to launch successful spoofing and tampering attacks without the explicit knowledge of the Data-ID, and the rules to derive the secret key for a particular PDU.

As shown in Figures 6.8 and 6.9, the SecOC verification mechanism works by comparing the authenticator value (MAC), and the freshness value (FV) of the incoming secured PDU data with the values calculated/obtained by the receiver itself. Because both sender and receiver know the secret key, as well as the other parameters to correctly calculate the MAC for each PDU, the receiver can perform the same MAC calculation and compare its own-calculated with the received one. The receiver also needs to check the validity of the incoming FV to prevent replay attacks.

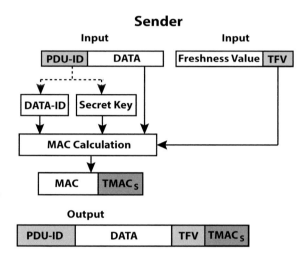

Figure 6.8: Secured PDU generation mechanism. The MAC of the data payload is calculated using some part of the Freshness Value (Truncated Freshness Value/TFV) as well as a DATA-ID and a secret key unique to the PDU-ID. Finally, the sender transmits the TFV and some part of the MAC value (Truncated MAC/TMAC) along with the original PDU-ID and data payload to the receiver as the Secured PDU. (Adapted from "Requirements on Secure Onboard Communication AUTOSAR CP Release 4.3.1", by Automotive Open System Architecture (AUTOSAR), 2017, p. 30)

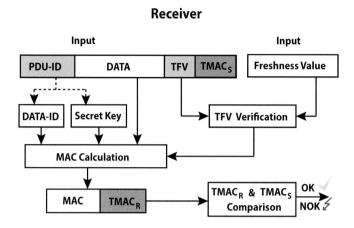

Figure 6.9: Secured PDU verification mechanism. The receiver validates the TFV of the Secured PDU by comparing its value with the actual Freshness Value. The receiver performs the same MAC calculation based on the DATA-ID and the secret key derived from the PDU-ID and compares its own calculation's result with the received TMAC. Only if both TFV and TMAC values are valid, can the receiver trust the data payload within the Secured PDU. Note that throughout the whole generation/verification process the data payload is transmitted as 'plain' data, i.e., without any encryption or any modification. (Adapted from "Requirements on Secure Onboard Communication AUTOSAR CP Release 4.3.1", by Automotive Open System Architecture (AUTOSAR), 2017, p. 30)

In addition to the above security mechanisms, AUTOSAR also provides ways for SWCs (software components) to interact directly with the SecOC module, such as to let the SWC get notified on a certain verification status, or to temporarily or permanently override the SecOC verification status [5]. Hence, the application might be able to detect a potential intrusion, e.g., when the number of failed verifications grows, or to take some preventative actions, such as ignoring certain suspicious PDUs.

6.2.3.4 Secure external communication

Up until now, we have discussed in-vehicle security or *embedded security*. Embedded security is essential, but as SDV functionality depends increasingly on information or services from outside the vehicle, secure external communication is just as important from both safety and privacy viewpoints. SDV external communication can take place with manufacturer's or operator's back-end servers, or with other vehicles, roads infrastructure, etc. also known as Vehicle-to-Everything (V2X) communication, which we will discuss later in this chapter.

External communication between the SDV and back-end servers is similar to general client/server communication in the classical IT world, so that many standard and well-proven methodologies or best-practices for security and privacy from the IT world can also be applied here as well. This also implies that security and privacy depend highly on the quality and effectiveness of the security mechanisms enforced by each SDV manufacturer. In contrast to the proprietary solutions used for communication with back-end servers, V2X communication works on the basis of standards-compliant communication protocols. Hence, security standards for V2X communication are necessary to ensure security without compromising interoperability.

Historically, V2X communication, or Intelligent Transportation Systems (ITS) in general, were researched by many public and private institutions in parallel. This led to several competing security proposals as well as standardization initiatives by various standardization bodies around the globe. Among them are the Security Credential Management System (SCMS), developed by the US Department of Transportation (USDOT) [44] and a series of standards published by the European Telecommunications Standards Institute (ETSI) [10].

Despite differences in architecture and technical details in these standards, secure V2X communication generally employs a *Public Key Infrastructure (PKI)* to facilitate credential verification of all communication partners, as well as to maintain trust relationship between authorities in multiple V2X networks. Proof of identity is usually conducted by verifying the *digital certificate* of the communication partner. The digital certificate is issued by an independent entity called the *certificate authority (CA)*. The primary role of CAs is to act as a trusted entity, which confirms that the holders of the issued digital certificate really are who they claim to be. Another task of CAs is to maintain the *Certificate Revocation List (CRL)*, which lists all certificates that ought not be trusted, despite their active validity period. With the help of public-key cryptography, and one-way hash functions, the digital certificate of the other communication partner can be authenticated before initiating the secure communication. Detailed explanations of PKI, public-key cryptography and certificate management are beyond the scope of this book. Interested readers are referred to technical books dedicated on this topic, such as [2] or [7].

The ISO/IEC 15408 security standard defines four privacy aspects that need to be protected from discovery and identity misuse [21]: *anonymity, pseudonymity, unlinkability* and *unobservability*. Anonymity means that the identity of users, who interact with the system or make use of a service, cannot be determined. Pseudonymity implies that a user can still be held accountable for system interactions or service usage; however, the real user identity remains undisclosed. Unlinkability means other entities are unable to determine whether multiple system interactions or multiple service usages are caused by the same user. Finally, unobservability is associated with the ability to interact with a system or use a service without other entities knowing that the system/service is being used.

In order to protect privacy in V2X network communications, ETSI ITS standards recommend PKI with separated CAs (see Figure 6.10) for credential authentication and service authorization [11]. The *Enrollment Authority (EA)* facilitates the identity verification of communication participants (users) using public-key cryptography and digital certificates as explained above. On valid authentication, the EA issues a temporary identity in the form of a *pseudonymous certificate*, also known as *enrollment credentials (EC)*. In order for a user to use V2X network service, the user needs to request permission from the *Authorization Authority (AA)* by sending its EC. Upon successful verification of the EC, the AA issues the authorization certificate or *authorization ticket (AT)* for each requested service. Proof for EA and AA credentials is provided by the *Root Authority (RA)*, the highest CA in the authority hierarchy. All involved CAs (RA, EA and AA) monitor the issued certificates, and maintain their own CRLs.

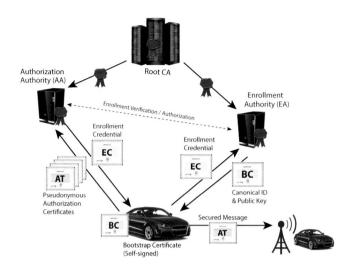

Figure 6.10: ETSI V2X PKI architecture. (Adapted from "Intelligent Transport System (ITS); Security; Pre-standardization study on pseudonym change management", by European Telecommunications Standards Institute (ETSI), 2018, ETSI TR 103 415 V1.1.1 (2018-04), p. 15. ©2018 ETSI)

6.2.4   Challenges

Linus Torvalds, the inventor of the Linux operating system, once stated, 'a truly secure computing platform will never exist' and argued that the most secure system might not even be usable [30]. Therefore, designing a secure SDV is always associated with finding a reasonable trade-off between security and other important aspects, such as cost, performance, and comfort. Cybersecurity criminals will always try to discover and exploit new vulnerabilities, while security experts try to close security holes as they are exploited. More connectivity also means more potential cybersecurity targets. Hence cybersecurity seems a never ending 'battle'.

As mentioned in the beginning of this section, the high complexity of any SDV system, coupled with dependency on third-party *black box* software, make it challenging to enforce the end-to-end security of the whole technology stack. Hence, professional security testing, e.g., penetration testing (*pen testing*), and independent security audit/review should be an integral part of the overall security strategy.

However, the biggest challenge is sometimes the (lack of) *security and safety culture* that seem to plague many tech companies worldwide. Not only the lack of awareness/enforcement of security best practices by the developers, but also economics, fierce competition and pressure from investors, which mean *visible* or *demonstrable* features are often prioritized higher than *invisible* security and safety features or fundamental architecture work. Since some safety and security-related decisions cannot simply be changed or implemented later as *add-ons*, this approach can lead to products with poor security and/or with higher overall development and maintenance cost.

6.3 Vehicle-to-everything communication

Communication with back-end servers helps SDVs to function more reliably beyond the boundary of their (limited) perception capabilities. However, the global information sent by these back-end servers might not always be sufficient to assist individual SDVs within a large scale deployment, and therefore needs to be augmented by locally relevant information. Furthermore, the proprietary nature of the back-end communication means the information is usually only available to vehicles from the same manufacturer, which makes the sharing of information between vehicles from different manufacturers a challenging task. This is where Vehicle-to-Everything, or V2X, communication comes in.

V2X is a general term for external communication between vehicles and other participants, mobile or stationary, in an intelligent traffic system. ETSI distinguishes four types of V2X communication: *Vehicle-to-Vehicle (V2V)*, *Vehicle-to-Infrastructure (V2I)*, *Vehicle-to-Network (V2N)*, and *Vehicle-to-Pedestrian (V2P)* [12]. V2V focuses on information exchange between vehicles in close proximity to each other. V2I refers to direct communication between vehicles and intelligent roads infrastructure, also known as Roadside Units (RSUs). V2N is associated with communication between vehicles and the Internet. V2P encompasses vehicle communication with human traffic participants, e.g., pedestrians, cyclists, etc.

6.3.1 Why is V2X important?

The ultimate objective of V2X is to improve traffic safety by helping vehicles and other traffic participants to avoid accidents. The information obtained from V2X communications can be processed and presented as warnings to human drivers, or it can trigger some safety mechanism in automated vehicles. V2X enables SDVs to perceive critical situations beyond their line-of-sight and internal information, e.g., a temporary change of lane direction, or lanes closed due to construction or accidents, or when the perception capability is temporarily limited, e.g., due to sudden bad weather conditions.

Another important V2X goal is to increase efficiency, particularly in the context of traffic and energy efficiency. With the help of accurate and locally relevant V2X information, vehicles can be informed to take alternative routes to reduce traffic congestion, or to adjust its speed to improve the overall traffic flow. The dynamic information can also be used by smart vehicles to optimize their energy consumption or reduce the environmental cost associated with driving.

6.3.2 V2X standards

Taking full advantage of V2X requires the active involvement of as many participants as possible. Thus, standardization is required to ensure interoperability between devices (or stations) from various manufacturers. Ironically, V2X has diverged into two competing incompatible standards, namely the IEEE 802.11p-based standards and the more recently 3GPP Cellular Vehicle-to-Everything (C-V2X) standards. The IEEE 802.11p standard is an amendment to the popular IEEE 802.11 Wi-Fi family of standards that enables the use of Wi-Fi for vehicular network communication. Both the U.S. Dedicated Short Range Communication (DSRC) as shown in Figure 6.11, and the European ETSI ITS-G5 V2X (see Figure 6.12) standards use IEEE 802.11p for the physical (PHY) and medium access control (MAC) layers in their protocol stacks [1]. The C-V2X standards were developed by the 5G Automotive Association (5GAA), a consortium of companies from automotive, technology and telecommunication industries, founded in 2016. C-V2X standards use cellular networks (LTE-4G and 5G) for V2X communication. This means the physical layer is fully incompatible with IEEE 802.11p. However, C-V2X reuses higher layer protocols and services from the DSRC and ITS-G5 standards.

All V2X standards have different protocol stacks for safety-critical and non-safety-critical applications. Non-safety-critical applications typically use TCP/UDP and IPv6 for transport layer and network layer protocols, respectively. The transport and network protocols for safety-critical applications differ from one standard to another: DSRC uses IEEE 1609.3 WAVE Short Message Protocol (WSMP) [20], whereas ITS-G5 uses Basic Transport Protocol (BTP) [14] and the GeoNetworking protocol [13]. Based on European standards, safety-critical messages between V2I participants are exchanged as *Cooperative Awareness Message (CAM)* and the *Decentralized Environment Notification Message (DENM)*. CAM is sent periodically between 10 Hz and 1 Hz and provides status information, e.g., heading, speed, lane position, etc. to other participants in the V2X network. DENM is an event-driven message, which is triggered whenever a triggering condition is fulfilled, e.g., detection of a traffic jam, and is repeatedly transmitted until the terminating condition is reached, e.g., end of traffic jam. The DSRC-based V2X network uses a set of safety messages defined in the SAE J2735 standard, such as the Signal Phase and Timing (SPaT) message for communicating traffic light status and the Traveler Information Message (TIM) for exchanging road conditions or other relevant information, e.g., road-work zone or the recommended speed for a particular road section [38].

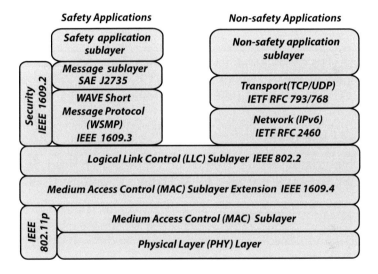

Figure 6.11: DSRC Protocol Stack. (Adapted from "Hybrid Adaptive Beaconing in Vehicular Ad Hoc Networks: A Survey", by Safdar Bouk, Kim Gwanghyeon, Syed Hassan Ahmed, Dongkyun Kim, 2015, doi: 10.1155/2015/390360, International Journal of Distributed Sensor Networks, 02, p. 16)

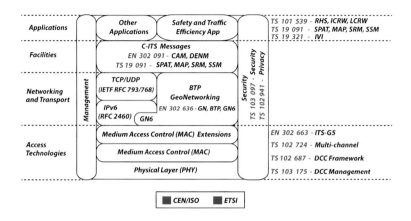

Figure 6.12: ITS-G5 Protocol Stack. (Adapted from "Cooperative intelligent transport systems standards in Europe", by Andreas Festag, 2014, doi: 10.1109/MCOM.2014.6979970, IEEE Communications Magazine, 52, p. 166-172)

6.3.3 V2I use cases

V2I involves communication between vehicles and Road Side Units (RSUs). Some example use cases for V2I communications are as follows:

6.3.3.1 Road work warning

Road works undoubtedly create one of the most difficult driving scenarios for SDVs. What makes it so challenging is that it requires an accurate dynamic interpretation of the new driving environment in a timely manner. In other words, the internal beliefs, or information stored in the vehicle's memory, cannot be depended upon: lane closures, changes in driving lane geometry and lane direction, unclear lane boundary markings (often both old and new markings are still visible), changes of speed limit, new traffic signs, closer distance to humans, other vehicles or road boundaries, etc. So, warning SDVs about road works ahead is clearly very useful, and allows them to prepare by adjusting their internal system to the new road conditions or by calculating an alternative route to avoid the road works completely. An example road work warning is shown in Figure 6.13.

Figure 6.13: Example of V2I road construction ahead warning.

6.3.3.2 Road hazard and accident warning

Road hazard refers to anything on the road surface that may endanger driving. It can be road debris, animals, ice, etc. Early warning of road hazards and accidents, e.g., with the help of DENM messages, alerts vehicles to search for alternative routes to prevent potential and subsequent accidents or to reduce the ongoing traffic congestions. The warning, as illustrated in Figure 6.14, is also a useful trigger for SDVs to increase the level of system awareness, as not all road hazards might be easily perceivable by the sensors (e.g., ice).

Figure 6.14: Example of a road hazard warning.

6.3.3.3 *Traffic light phase event*

Information about the traffic light phase, and the residual time for the current phase, helps improve safety and efficiency. Traffic light phase recognition using vehicle sensors, e.g., cameras, might not always work reliablely in all traffic situations or weather conditions, so the traffic light phase information can be used to complement the vehicle's perception. Using Signal Phase and Timing (SPaT) and MAP messages, both defined in SAE J2735 standard [38], allows the current phase of traffic light for each lane, the residual time for the current phase, and the physical geometry of the intersections to be exchanged over V2I networks. Figure 6.15 shows an example application of V2I traffic light phase event.

Figure 6.15: Example of a traffic light phase event. (Reprinted with permission from Continental AG. ©2017 Continental AG)

6.3.4 *V2V use cases*

In the following section, we will cover some examples of V2V use cases.

6.3.4.1 Intersection movement assist warning

Based on US National Highway Traffic Safety Administration (NHTSA) analysis, about 40% of traffic accidents that happened in the US in 2008 were related to intersections [8]. Furthermore, the publication also identified 'turned with obstructed view' as the major cause for intersection-related accidents. Due to limited line-of-sight, and physical sensor limitations in general, it is unclear whether SDVs using on-board sensors alone can improve the statistics, i.e., reduce the rate of intersection-related accidents. Intersection movement assist warning, as illustrated in Figure 6.16, helps prevent accidents by providing detailed movement information to other road participants around the intersection, and therefore extends the vehicles' perception capabilities beyond their line-of-sight.

Figure 6.16: Example of a pedestrian warning on intersection. (Reprinted with permission from Continental AG. ©2017 Continental AG)

6.3.4.2 Wrong way driver warning

Wrong way driving refers to situations where a vehicle is driving in the opposite direction for the lane it is in, e.g., it is driving straight at other vehicles. Wrong way driving is especially dangerous on highways because other vehicles are moving with high speed and therefore stopping distances are increased. Also, on a highway there may be less scope for taking avoiding action. Sending timely warnings of the presence of a wrong way driver, as shown in Figure 6.17, gives other vehicles a chance to react and thus helps prevent accidents.

Figure 6.17: Example of a wrong way warning.

6.3.4.3   Do-not-pass warning

According to a factsheet published by the British Royal Society for the Prevention of Accidents (RoSPA), overtaking is regarded as one of the highest risk driving maneuvers, and overtaking a moving vehicle on the offside accounted for over half of all overtaking-related accidents in 2015 [35]. Overtaking involves a very complex estimation and decision-making process in human brains, as well as well-coordinated actions during preparation, execution, and termination of the maneuver. If the driver has failed to spot an unsafe condition, such as a blind bend or a slow vehicle in the overtaking lane, do-not-pass warnings (see Figure 6.18) can be life-saving information. For SDVs, such warnings are extremely useful as they effectively extend the range of the vehicle's own sensors.

Figure 6.18: Example of a do-not-pass warning.

6.3.5 V2P use case

The following example shows how V2P can be used.

6.3.5.1 Vulnerable road user warning

In general, drivers tend to assume pedestrians are able to see and hear them, and thus will be aware of their presence. However, some pedestrians may have impaired senses, or may be temporarily distracted. This can make it harder for them to be aware of vehicles when crossing the road. This is especially the case at intersections or zebra crossings where pedestrians may assume it is safe to cross. A Vulnerable Road User (VRU) warning can be used to alert vehicles that a pedestrian with reduced mobility or impaired senses is crossing or is about to cross the road. For SDVs, this means they are not solely dependent on things like lidar and cameras to detect pedestrians.

Similarly, VRU can be used to alert vehicles to the presence of cyclists. Cyclists are significantly more vulnerable than people in vehicles, because they have little or no physical protection in an accident. Also, even with good lights, cyclists are often very hard to spot at night, since they tend to be close to the kerb, and can merge into their background. Using VRU to warn about the presence of a cyclist will be especially useful for SDVs, since it will augment their own ability to detect the cyclist.

6.3.6 Challenges

One of the main V2X challenges is network coverage. Typically, V2X networks are only available for certain geographical areas because they involve high investment (and sometimes bureaucracy) from public road authorities or municipalities to upgrade the existing road infrastructure with roadside units (RSU), intelligent sensors, communication backbones, etc. Also, as of writing, the majority of vehicles worldwide do not have V2X support, as this is not mandatory. In late 2016, the US NHTSA proposed a mandate that would require all new light vehicles to support DSRC-based V2V communication [32]. However, the future of the mandate seems unclear at the moment [28].

Another big challenge is the issue of interoperability. Historically, standards for V2X communication were developed in parallel by different standardization bodies throughout the globe, particularly in the US and Europe. The parallel standardization initiatives have led to a set of different standards that are not fully compatible with each other. This is similar to the situation that existed in the early days of digital cell phones (even now the US uses a different set of cellular frequencies to most of the rest of the world). In November 2009, representatives of the European Commission and USDOT signed the 'Joint Declaration of Intent on Research Cooperation in Cooperative Systems' that marked the beginning of harmonization process between US and European ITS standards [43]. However, the harmonized standards may not be sufficient to converge the whole V2X ecosystem, since the main reason for the current V2X fragmentation is actually the existence of the competing C-V2X standards.

6.4 Back-end systems

Even with the most sophisticated hardware and intelligent software installed, there is still a limit to what an SDV is able to perceive externally. The limitation may be caused by the maximum working range of the sensor, sensor occlusion, poor weather, etc. So, it might sometimes be essential for SDVs to get some external information beyond what they can perceive themselves, so they can plan ahead and make better decisions in a more timely manner. Therefore, SDVs usually operate in conjunction with some back-end services provided by the operator or manufacturer.

6.4.1 Why are back-end systems important?

As mentioned above, one of the most important reasons why back-end systems are needed is that SDV perception is somewhat limited and the systems can help provide additional information beyond their 'line of sight' that might be relevant for the current or near-future driving situations. Vehicle-to-Vehicle (V2V) or Vehicle-to-Infrastructure (V2I) communication can also help with this; however, these are only available if the SDV is within working proximity of other V2V-capable vehicles or V2I-supporting infrastructure.

Another important reason is the need for real-time traffic information or live map updates. Some internal information stored in the vehicle might not be accurate anymore, or may need to be updated in real-time. One example is the temporary closure of a lane, or creation of a contraflow due to an accident or construction site. The back-end servers need to aggregate information from various sources, and pass the relevant information to each SDV.

Depending on the computing resource available, some heavy or resource-intensive operations might need to be offloaded to the back-end systems, and the computation result sent back to the SDV. This approach is common in mobile phones. Typically, the operations involved are non-safety-critical computations that need to process a lot of data, operations that use data not locally available to SDVs, or functions that can be more efficiently performed on the back-end servers.

Back-end systems are also essential for fleet management tasks. Take autonomous public shuttles as an example. The back-end system might automatically reschedule a route if there are roadworks, or direct certain shuttles to charging stations based on the real-time situation. Also, reserve shuttles might be dispatched automatically in time of high demand. SDV operators or manufacturers might also need to perform remote diagnostics, real-time monitoring of a vehicle's status, or trigger *Over-The-Air* (OTA) software update through the back-end systems.

6.4.2 Back-end system functions

Let us have a closer look at some of the aforementioned back-end system functions.

6.4.2.1 Software over-the-air update

The 'traditional' or conventional way of updating vehicle software is through a special tool attached to the vehicle during a visit to a repair shop. Software Over-The-Air (SOTA) is a way to perform vehicle software updates remotely using common communication networks, e.g., public cellular LTE networks or private/public Wi-Fi hotspots. SOTA is the established way of conducting software updates for consumer electronics, and car manufacturers are embracing the SOTA idea and working actively on supporting SOTA as an alternative software update mechanism.

6.4.2.2   High-definition maps

As we saw in Chapter 3, an SDV localizes itself using maps. In order to localize accurately, SDVs need High-Definition (HD) maps, i.e., maps of the operating environment with very high precision (often just centimeters). HD maps, as shown in Figure 6.19, however, require a huge amount of data to be stored. Depending on the available hardware resources of the SDV computing platform and the size of the SDV operational area, it might not be possible to store all the map data in the vehicle. Thus, SDVs can request new or missing map data on-demand from the back-end server. The back-end server can also update the internal map or trigger notifications for relevant events along the route, such as lane closures due to road works, traffic accidents, etc.

Figure 6.19: Example of a high-definition map. (Adapted from "Simultaneous Localization and Map Change Update for the High Definition Map-Based Autonomous Driving Car", by Kichun Jo, Chansoo Kim, Myoungho Sunwoo, 2018, Sensors 2018, 18(9):3145. ©2018 Kichun Jo, Chansoo Kim, Myoungho Sunwoo. CC BY 4.0)

6.4.2.3 Fleet management

In some SDV use cases, such as autonomous public shuttles or last mile delivery systems, multiple SDVs operate together to provide a service. Fleet management services ensure smooth, safe and efficient operation of all the SDVs. Typical fleet management services include location tracking of each SDV, service dispatching, dynamic route calculation, system health monitoring, and remote diagnostics as shown in Figure 6.20. The fleet management services might be performed manually by human operators in a control room, automated using fleet management software on the back-end servers, or a combination of both.

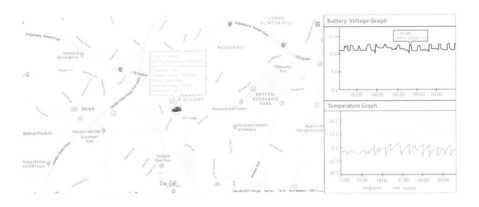

Figure 6.20: Example of a fleet management software.

6.4.3 Challenges

Although the purpose of back-end systems is to ensure that SDVs drive smoothly, safely, and reliably, there are some aspects that need to be addressed carefully before adding back-end systems to the overall SDV system equation. The first aspect is cybersecurity, which involves securing the system against external attacks and ensuring data privacy. The overall system security is effectively determined by how well the manufacturers or operators secure their external communications. Another aspect is the Quality-of-Service (QoS) of the communication network used, e.g., the latency, throughput, and loss. Depending on how critical the data exchange between SDVs and the back-end servers is, using private dedicated wireless networks might be better than public cellular ones, but it would imply significantly higher installation and maintenance costs.

The proprietary nature of the communication protocols between SDVs and back-end systems gives manufacturers (or OEMs) full control over what kind of data can be exchanged and how often. Although this may sound advantageous from the privacy perspective (since the data are not shared with other parties), it does restrict access to potentially useful information from other road participants. Another challenge is related to the coverage of the individual back-end services. Due to the sheer amount of data, an HD map might only be available in certain geographical locations. Thus, an SDV may face difficulties localizing in a new area if the corresponding HD map for that area is obsolete or not yet available.

6.5   Summary

This chapter looked at some of the external factors that have to be considered when designing an SDV. Chief among these are the related issues of safety and security. We also looked at the importance of external data to SDVs, including data from fleet management systems and V2X networks.

As we saw, functional safety is a critical aspect for all vehicle manufacturers, but especially for SDV manufacturers. The aim of functional safety standard such as ISO 26262 is to reduce the risk of harm that might be caused by the failure of one or more systems or components. In the automotive industry, these risks are classified according to the Automotive Safety Integrity Level (ASIL), where ASIL D reflects the most severe class of events. Functional safety is a continuous process that happens throughout a vehicle's lifetime, from the start of the design process through to decommissioning and dismantling the vehicle. Every risk that is identified during the design phase has to be mitigated using a suitable approach (or combination of approaches) for the level of risk.

Another important technology aspect is cybersecurity, or protecting the computing and network systems from external attacks. As explained, even vehicles that are not permanently connected to an external network can be hacked, for instance by injecting a firmware update that opens up the CAN bus, and allows fake messages to be passed to the vehicle. These risks are far higher when there is permanent or regular external connectivity, such as V2X. As a result, much work has been done on cybersecurity standards for vehicles, resulting in the publication of ISO/SAE 21434. We looked at some of the approaches that have been taken to mitigate the risk of cybersecurity attacks, including the use of HSMs and secure network communications. We saw a detailed description of the different levels of EVITA HSM, and showed how HSMs can be used to secure communications between vehicle sensors and other modules. We also explored the AUTOSAR SecOC, which is the AUTOSAR recommended approach for providing secure in-vehicle network communications. At the end of the section we looked at how PKI-based systems can be used to provide a suitable level of security for communications between the vehicle and the outside world based on certificates and trust.

Next we looked in detail at the different forms of V2X network. We saw examples of how these networks can be used to help better inform both drivers and SDVs, and looked at the challenges that make them hard to install at a large scale. One of the key forms of V2X is vehicle to infrastructure, or V2I. This can enable useful functions including road work warning, road hazard and accident warning, and traffic light phase information. V2V (vehicle-to-vehicle) warnings can be used to alert other vehicles not to overtake, or to warn of the presence of wrong-way vehicles. Vehicle-to-Person (V2P) is primarily focussed on warning vehicles of the presence of pedestrians and cyclists either in the road or about to cross the road. Pedestrians and cyclists are the most vulnerable road users, and hence anything that can help protect them is vital.

At the end of the chapter, we discussed how back-end systems can be used to assist SDVs. Live map updates can be used both to ensure that SDVs are aware of any changes to road layouts (be they temporary or permanent), and to provide dynamic high-definition maps with centimeter accuracy for highly accurate localization. Fleet management systems allow operators of large vehicle fleets to perform tasks like dynamic scheduling, vehicle tracking, and can even be used to monitor the mechanical health of the fleet. Finally, SOTA allows manufacturers and operators to update a vehicle's software remotely. For SDVs, running the latest software could be critical, especially given the pace of software development, and the likely improvements in performance that might bring.

In the next chapter, we will discuss some of the proposed use cases for SDVs, along with a look at alternative approaches that might speed up SDV development, such as deep learning.

References

[1] Khadige Abboud, Hassan Aboubakr Omar, and Weihua Zhuang. Interworking of DSRC and cellular network technologies for V2X communications: A survey. *IEEE transactions on vehicular technology*, 65(12):9457–9470, 2016.

[2] Carlisle Adams and Steve Lloyd. *Understanding PKI: concepts, standards, and deployment considerations*. Addison-Wesley Professional, 2003.

[3] AUTOSAR. Requirements on secure onboard communication. https://www.autosar.org/fileadmin/user_upload/standards/classic/4-3/AUTOSAR_SWS_SecureOnboardCommunication.pdf. [accessed 20-May-2018].

[4] AUTOSAR. Specification of crypto service manager. https://www.autosar.org/fileadmin/user_upload/standards/classic/4-3/AUTOSAR_SWS_CryptoServiceManager.pdf. [accessed 20-May-2018].

[5] AUTOSAR. Specification of secure onboard communication. https://www.autosar.org/fileadmin/user_upload/standards/classic/4-3/AUTOSAR_SWS_SecureOnboardCommunication.pdf. [accessed 20-May-2018].

[6] BMJV. Gesetz über die haftung für fehlerhafte produkte (produkthaftungsgesetz - prodhaftg). https://www.gesetze-im-internet.de/prodhaftg/ProdHaftG.pdf. [accessed 20-May-2018].

[7] Johannes A Buchmann, Evangelos Karatsiolis, and Alexander Wiesmaier. *Introduction to public key infrastructures*. Springer Science & Business Media, 2013.

[8] Eun-Ha Choi. Crash factors in intersection-related crashes: An on-scene perspective. Technical report, 2010.

[9] Morris J Dworkin. Recommendation for block cipher modes of operation: The CMAC mode for authentication. Technical report, 2016.

[10] ETSI. Automotive intelligent transport systems. `https://www.etsi.org/technologies-clusters/technologies/automotive-intelligent-transport`. [accessed 20-May-2018].

[11] ETSI. Etsi - ts 102 941. Intelligent transport systems (ITS); security; trust and privacy management. `https://www.etsi.org/deliver/etsi_ts/102900_102999/102941/01.02.01_60/ts_102941v010201p.pdf`. [accessed 20-May-2018].

[12] ETSI. Etsi - ts 122 185. Requirements for V2X services. `https://www.etsi.org/deliver/etsi_ts/122100_122199/122185/14.03.00_60/ts_122185v140300p.pdf`. [accessed 20-May-2018].

[13] ETSI. Ts 102 636-4-1 v1.2. Intelligent transport systems (ITS); vehicular communications; geonetworking; part 4: Geographical addressing and forwarding for point-to-point and point-to-multipoint communications; sub-part 1: Media-independent functionality.

[14] ETSI. Ts 102 636-5-1 v1.2. Intelligent transport systems (ITS); vehicular communications; geonetworking; part 5: Transport protocols; sub-part 1: Basic transport protocol.

[15] Andy Greenberg. Hackers remotely kill a jeep on the highway - with me in it. *Wired*, 7:21, 2015.

[16] Trusted Computing Group. Tcg tpm 2.0 automotive thin profile for tpm family 2.0; level 0. `https://trustedcomputinggroup.org/resource/tcg-tpm-2-0-library-profile-for-automotive-thin/`. [accessed 20-May-2018].

[17] Tim Güneysu and Christof Paar. Ultra high performance ECC over NIST primes on commercial FPGAS. In *International Workshop on Cryptographic Hardware and Embedded Systems*, pages 62–78. Springer, 2008.

[18] Olaf Henniger, Alastair Ruddle, Hervé Seudié, Benjamin Weyl, Marko Wolf, and Thomas Wollinger. Securing vehicular on-board IT systems: The EVITA project. In *VDI/VW Automotive Security Conference*, 2009.

[19] IEC. Functional safety and iec 61508. https://www.iec.ch/functionalsafety/. [accessed 20-May-2018].

[20] IEEE. IEEE standard for wireless access in vehicular environments (wave)-networking services. *IEEE 1609 Working Group and others*, pages 1609–3, 2016.

[21] ISO. ISO - international organization for standardization. information technology - security techniques - evaluation criteria for it security - part 2: Security functional requirements. https://www.iso.org/standard/40613.html. [accessed 20-May-2018].

[22] ISO. Iso/sae cd 21434. road vehicles - cybersecurity engineering. https://www.iso.org/standard/70918.html. [accessed 20-May-2018].

[23] ISO. Road vehicles - functional safety - part 9: Automotive safety integrity level (ASIL)-oriented and safety-oriented analyses. https://www.iso.org/standard/51365.html. [accessed 20-May-2018].

[24] ISO. 26262: Road vehicles-functional safety. *International Standard ISO/FDIS*, 26262, 2011.

[25] Matthias Klauda, Stefan Kriso, Reinhold Hamann, and Michael Schaffert. Automotive safety und security aus sicht eines zulieferers, page 13, 2012.

[26] Loren Kohnfelder and Praerit Garg. The threats to our products. *Microsoft Interface, Microsoft Corporation*, 1999.

[27] Tim Leinmüller, Levente Buttyan, Jean-Pierre Hubaux, Frank Kargl, Rainer Kroh, Panagiotis Papadimitratos, Maxim Raya, and Elmar Schoch. Sevecom-secure vehicle communication. In *IST Mobile and Wireless Communication Summit*, number LCA-POSTER-2008-005, 2006.

[28] Joan Lowy. Apnewsbreak: Gov't won't pursue talking car mandate. https://apnews.com/9a605019eeba4ad2934741091105de42. [accessed 20-May-2018].

[29] Georg Macher, Harald Sporer, Reinhard Berlach, Eric Armengaud, and Christian Kreiner. Sahara: a security-aware hazard and risk analysis method. In *Proceedings of the 2015 Design, Automation & Test in Europe Conference & Exhibition*, pages 621–624. EDA Consortium, 2015.

[30] Andy Meek. Linux creator explains why a truly secure computing platform will never exist. https://bgr.com/2015/09/25/linus-torvalds-quotes-interview-linux-security/. [accessed 20-May-2018].

[31] Charlie Miller and Chris Valasek. Remote exploitation of an unaltered passenger vehicle. *Black Hat USA*, 2015:91, 2015.

[32] NHTSA. Federal motor vehicle safety standards; V2V communications. *Federal Register*, 82(8):3854–4019, 2017.

[33] Jonathan Petit, Bas Stottelaar, Michael Feiri, and Frank Kargl. Remote attacks on automated vehicles sensors: Experiments on camera and lidar. *Black Hat Europe*, 11:2015, 2015.

[34] Norbert Pramstaller, Christian Rechberger, and Vincent Rijmen. A compact FPGA implementation of the hash function whirlpool. In *Proceedings of the 2006 ACM/SIGDA 14th international symposium on Field programmable gate arrays*, pages 159–166. ACM, 2006.

[35] ROSPA. Road safety factsheet - overtaking. http://www.rospa.com/rospaweb/docs/advice-services/road-safety/drivers/overtaking.pdf. [accessed 20-May-2018].

[36] Deborah Russell, Debby Russell, GT Gangemi, Sr Gangemi, and GT Gangemi Sr. *Computer security basics*. O'Reilly Media, Inc., 1991.

[37] SAE. Cybersecurity guidebook for cyber-physical vehicle systems. http://standards.sae.org/j3061_201601. [accessed 20-May-2018].

[38] SAE. J2735: Dedicated short range communications (dsrc) message set dictionary. https://www.etsi.org/deliver/etsi_en/302600_302699/3026360401/01.02.00_20/en_3026360401v010200a.pdf. [accessed 20-May-2018].

[39] Christian Schleiffer, Marko Wolf, André Weimerskirch, and Lars Wolleschensky. Secure key management-a key feature for modern vehicle electronics. Technical report, SAE Technical Paper, 2013.

[40] Desmond Schmidt, Kenneth Radke, Seyit Camtepe, Ernest Foo, and Michał Ren. A survey and analysis of the GNSS spoofing threat and countermeasures. *ACM Computing Surveys (CSUR)*, 48(4):64, 2016.

[41] Martin Schmidt, Marcus Rau, Ekkehard Helmig, and Bernhard Bauer. Functional safety-dealing with independency, legal framework conditions and liability issues. *Official Journal of the European Union dated*, 50(200/1), 2009.

[42] Bernd Spanfelner, Detlev Richter, Susanne Ebel, Ulf Wilhelm, Wolfgang Branz, and Carsten Patz. Challenges in applying the ISO 26262 for driver assistance systems. *Tagung Fahrerassistenz, München*, 15(16):2012, 2012.

[43] US-DOT. Program TITSS. development activities. ITS standards program | development activities | international harmonization. https://www.standards.its.dot.gov/DevelopmentActivities/IntlHarmonization. [accessed 20-May-2018].

[44] William Whyte, André Weimerskirch, Virendra Kumar, and Thorsten Hehn. A security credential management system for V2V communications. In *VNC*, pages 1–8, 2013.

[45] Ulf Wilhelm, Susanne Ebel, and Alexander Weitzel. Funktionale sicherheit und iso 26262. In *Handbuch Fahrerassistenzsysteme*, pages 85–103. Springer, 2015.

[46] Marko Wolf and Timo Gendrullis. Design, implementation, and evaluation of a vehicular hardware security module. In *International Conference on Information Security and Cryptology*, pages 302–318. Springer, 2011.

[47] Stephen S. Wu. *Product Liability Issues in the U.S. and Associated Risk Management*, pages 575–592. Springer Berlin Heidelberg, Berlin, Heidelberg, 2015.

[48] Chen Yan, Wenyuan Xu, and Jianhao Liu. Can you trust autonomous vehicles: Contactless attacks against sensors of self-driving vehicle. *DEF CON*, 24, 2016.

Chapter 7

Applications and outlook

We have now met all the key components of SDVs, seen how they are built, and discussed some other related technology aspects. Huge technological efforts such as the space program and ongoing research have taken humans into space and given us things like GNSS, satellite communications, and more accurate weather forecasting. But they have also led to the development of many practical products that we use in our daily life, such as memory foam and solar cells. These *side innovations or spin-off technologies* originated from NASA's research [2]. Similarly, not only will research and development of SDV technology eventually bring us fully autonomous cars with all their expected benefits, but the long journey to achieve that goal will also generate a wide range of exciting innovations along the way.

We currently live in a very exciting time for SDVs, and we may witness one of the most disruptive innovations in personal mobility coming to fruition. But even though SDV technology has been evolving rapidly in recent years, many challenges still need to be solved. As more and more companies from outside the traditional automotive industry take part in the global race for SDV development, the SDV technology landscape has become increasingly heterogeneous, with many specialized companies trying to solve specific problems within the SDV technology stack. The diverse backgrounds and strengths of all these companies lead to a variety of development philosophies and strategies aiming to make SDVs a reality.

Below, we will see some current examples of where SDV technology is being applied, and we will take a closer look at two emerging trends: one related to the development strategy and the other related to the involvement of Artificial Intelligence (AI), particularly deep learning.

7.1   SDV technology applications

In the following sections, we will cover some expected applications for SDV technology, both in transportation and non-transportation use cases.

7.1.1   Transportation use cases

Clearly, transportation is the domain that is impacted most by SDV technology. The recent significant media attention and research and development (R&D) investments in this area have spawned numerous start-ups in this field. It has also brought many other non-traditional automotive companies, e.g., IT and technology companies, into the growing SDV ecosystem, where they compete or partner with established companies from the traditional automotive ecosystem. By taking a look into what these companies are doing, we can have a better assessment of what is possible with this technology.

7.1.1.1   Private passenger cars

Nowadays, almost any (if not every) study or publication about future personal mobility will feature at least one self-driving car. Self-driving passenger cars for private or shared rides (also known as *robotaxis*) are undoubtedly a disruptive technology that reshapes our thinking about cars; not only our behavior when being driven in such cars, but also related issues such as insurance business models, liability, and many others.

Even though most automakers have been pursuing full autonomy as part of their Advanced Driver Assistance System (ADAS) roadmap, the race to build self-driving cars has been sparked by the increasing competition from new players outside the traditional automotive industry. The new competitors include IT giants, such as Waymo (part of Google's parent, Alphabet Inc.) and Baidu, as well as other tech companies, such as Lyft and Uber. Their core competence in software and other key enabling areas, such as AI and fast-paced consumer product development, might give them a competitive edge to compete with the established companies that come from a tradition of mechanical/automotive engineering with relatively more limited software backgrounds. Figure 7.1 depicts an example of such self-driving passenger cars.

Figure 7.1: Self-driving car. (©Dllu, https://commons.wikimedia.org/wiki/ File:Waymo_Chrysler_Pacifica_in_Los_Altos,_2017.jpg, https://creativecomm ons.org/licenses/by-sa/4.0/legalcode)

7.1.1.2   Public shuttles

Autonomous public shuttles are arguably the SDV use case that is likely to reach the highest levels of automation (level 4 or even level 5) in the near future. These self-driving shuttles typically operate on private or public roads within a confined and controlled space. They also have defined routes, such as transporting people from a train station to the airport terminal, between buildings in a large campus, or from one attraction to another in a large theme park. The higher level of automation can be achieved more easily due to the significantly reduced complexity in comparison to private passenger cars for use on public roads. Because of the limited scope of operational area, the whole environment can be mapped very accurately to allow precise localization. Additionally, those shuttles typically travel slowly along pre-defined routes, and are usually allowed to stop in case of uncertainty and wait until the situation clears. An example of such autonomous shuttles is shown in Figure 7.2.

An autonomous public shuttle system is generally supported by a back-end system on the operator's side for fleet management, service dispatching, health monitoring, etc. The back-end system may be manually or semi-manually controlled by human operators, or fully automated with human operators only as backup. The back-end system might also coordinate auxiliary components other than the autonomous shuttles, such as displaying passenger information inside the vehicle, responding to service requests at stops, controlling the ticketing system, etc.

Figure 7.2: Driverless shuttle. (©Richard Huber, https://commons.wikimedia. org/wiki/File:Autonomes_Fahren_in_Bad_Birnbach.jpg, https://creativecomm ons.org/licenses/by-sa/4.0/legalcode)

7.1.1.3   Last mile delivery

Delivering goods from source to user creates a long and complex logistics chain. Last mile delivery refers to the last part of this chain, which involves the movement of goods from a local distribution center to the individual consumer at the final destination. According to a McKinsey report in 2016, last mile delivery accounts for 50% or more of the total package delivery cost [16]. There are several reasons why last mile delivery is the least efficient leg of the logistics and supply chain. Traffic congestion and lack of parking in urban areas, longer travel to remote areas, and repeated delivery attempts due to unavailability of recipients, to name just a few. Thus, many logistics companies are embracing SDV technology, which has the potential to make last mile delivery more efficient and less expensive.

Driverless parcel delivery vehicles and delivery drones are two prominent examples of SDV technology applications to mitigate the last mile delivery issues. Driverless delivery robots, as shown in Figure 7.3, load parcels from a distribution center and navigate autonomously to reach the customer's address. Although autonomous navigation in 3D aerial space poses different challenges to 2D ground space, both share common basic technologies, for example in the areas of perception and navigation.

Figure 7.3: Delivery robot. (©User: Ohpuu / Wikimedia Commons / CC-Zero)

7.1.1.4 Road freight

In addition to the last mile delivery, SDV technology also has the potential to improve safety and efficiency in another part of the transportation sector, namely, in the domain of road freight for long distance goods transportation. According to the EU transport statistical pocketbook published in 2018, the vast majority of freight is transported by road [6]. However, road freight operators in Europe (and also the U.S.) face an increasing shortage of truck drivers, which makes self-driving trucks, as shown in Figure 7.4, a promising solution for the road freight sector [15]. Nevertheless, the driver shortage in Europe will persist, even with the introduction of self-driving trucks, according to a projection in the same report.

Truck platooning, a method of grouping two or more trucks traveling in a convoy, is a popular automated driving application for road freight transportation. The front vehicle leads the travel, and the other vehicles follow automatically, maintaining a safe distance from one another. Besides giving better fuel efficiency and increased safety, truck platooning is believed to be able to improve traffic flow and road usage [8].

Figure 7.4: Cockpit view of a self-driving truck. (Reprinted with permission from Freight-Match. ©2018 Freight-Match)

7.1.2 Non-transportation use cases

In this section, we will cover some examples where SDV technology is a key enabler to innovations outside the transportation domain.

7.1.2.1 Driverless tractors

Farming is one of the oldest human occupations and is traditionally a labor-intensive occupation. Due to structural changes in the workforce in modern society, many farmers, especially in developed countries, are finding it hard to employ sufficient labor during harvest, as people tend to prefer higher paid jobs in other industries [24]. As a result, the agriculture industry is gradually moving towards modern farming, employing high-tech tools and autonomous machines, such as driverless tractors or autonomous harvesters, as shown in Figure 7.5, in order to compensate for the labor shortage, as well as to increase productivity.

Driverless tractors are not only equipped with agriculture-specific sensors, such as ground moisture level sensors, but also a range of other sensors, e.g., GNSS receiver, radar or camera. With the help of these sensors, the tractors can perform basic SDV tasks, such as object detection for collision avoidance or localization for navigation throughout the field.

Figure 7.5: Two autonomous tractors performing a spraying function. (©ASIrobots, https://en.wikipedia.org/wiki/File:Autonomous_compact_ tractors_in_a_Texas_vineyard,_Nov_2012.jpg, "Autonomous compact tractors in a Texas vineyard, Nov 2012", https://creativecommons.org/licenses/by-sa/3.0/legalcode)

7.1.2.2 Emergency-response robots

On March 11, 2011, one of the most powerful earthquakes since the beginning of the 20th century hit the Pacific coast of Japan and triggered devastating tsunamis. This caused a significant death toll as well as extensive property damage. The disaster also caused a series of catastrophic accidents, most significantly the nuclear reactor meltdowns in the Fukushima region. These secondary events put humans and the environment at great risk because of the possible exposure to radioactive substances. Emergency-response robots are designed to aid humans to reach such disaster areas that are inaccessible or unsafe for humans. The robots need to cope with navigating through unknown harsh environments with limited or non-existent communication infrastructure. However, the magnitude of the Fukushima nuclear disaster, particularly the high levels of radiation, brought even the most advanced robots at that time to their limits [19].

The 2011 Fukushima disaster has inspired the robotics community to accelerate development in this field. The European Robotics League Emergency (formerly euRathlon) is the European competition for ground, underwater, and flying emergency-response robots using realistic mock scenarios inspired by the Fukushima accident [25]. Also inspired by the Fukushima disaster, the U.S. Defense Advanced Research Projects Agency (DARPA) initiated the DARPA Robotics Challenge (DRC) for designing robots that are capable of performing human tasks in emergency situations, such as driving a vehicle, walking through rubble and climbing stairs [7]. Figure 7.6 shows one of the participating robots at DRC 2015.

Figure 7.6: An emergency-response robot is turning a valve 360 degrees at the 2015 DARPA Robotics Challenge. (©Office of Naval Research from Arlington USA, https://commons.wikimedia.org/wiki/File:150605-N-PO203-329_(18529371672).jpg, "150605-N-PO203-329 (18529371672)", https://creativecommons.org/licenses/by/2.0/legalcode)

7.1.2.3 Security robots

Another application of SDV technology can be found in security robots. Security robots are designed to complement human guards by performing routine surveillance of large areas more efficiently and more reliably, regardless of weather conditions. Furthermore, security robots can be deployed to monitor hazardous or less accessible areas, and also provide an alternative security solution for areas where security camera installations are not feasible. An example of a commercial indoor security robot is shown in Figure 7.7.

One of the most important capabilities for security robots is reliable *anomaly detection*. In the event of an anomaly, the security robot alerts human guards or operators to take further actions. Consequently, too many *false positives*, i.e., false alarms, and *false negatives*, i.e., undetected anomalies, will significantly diminish the usefulness of such robots.

Figure 7.7: Security robot. (Reprinted with permission from Robot Robots Company. ©2016 RRC Robotics)

7.2 Trends in SDV development strategy

According to Beiker [3], there are three major development strategies employed by SDV developers: evolutionary, revolutionary, and transformative. Figure 7.8 shows some key market players within the SDV technology ecosystem. Which development strategy a particular SDV market player pursues depends on several factors, such as their key competencies, their main motivation, and their primary use case. In this section we will take a look at each of the three approaches, and the people behind them, to get a better understanding of how SDV technology is being developed.

Figure 7.8: SDV technology landscape. (Reprinted with permission from Phil Magney, VSI Labs. ©2019 VSI Labs)

7.2.1 Evolutionary

The proponents of the evolutionary development strategy are generally the major players in the automotive industry (OEMs and suppliers) who believe that fully autonomous driving can only be achieved through incremental improvement of the existing Advanced Driver Assistance Systems (ADAS).

Year after year, established companies in the automotive industry try to stay ahead of the competition and to attract new customers by introducing new innovations, ranging from seamless integration with consumer electronics to a whole range of new ADAS features. New ADAS innovations are built upon existing ADAS features with increasingly higher levels of automation. Thus, full autonomous driving is the logical endpoint of this evolutionary process.

Since the core business of the automotive industry is selling cars, SDVs need to function well as a product (and thus be sellable) in as many markets/regions as possible. Also, due to mass production and economies of scale, the less customizations required for SDVs to operate globally, the more costs can be saved. Hence, SDVs developed according to this strategy may start with relatively low levels of automation compared to the other strategies, but they are designed to cope in larger and more diverse geographical locations.

7.2.2 Revolutionary

The proponents of the revolutionary strategy argue that the evolutionary approach may well take too long to achieve full autonomy. The revolutionary strategy takes the position that the full autonomy can only be achieved through a *disruptive leap*, a paradigm change compared to how vehicles are developed by the automotive industry. The companies pursuing this strategy are usually IT or technology companies coming from outside the traditional automotive industry, such as Waymo. Thus, the incremental approach makes little sense, and would set them back a couple of years behind the competition.

There are some common characteristics shared by the companies behind the revolutionary strategy. Firstly, these companies have strong expertise in building software-based or data-driven products in short development cycles. Secondly, they typically have strong experience in Artificial Intelligence (AI). Hence, the revolutionary strategy backers build SDV based on their expertise in these areas. In other words, an SDV is primarily regarded as a software-, data-, and AI-driven product deployed on wheels. This philosophy is radically different to the common approach in the automotive industry, which emphasizes building safe, efficient and comfortable vehicles and adds more and more intelligence (automation) to them. The revolutionary strategy also leads to an unconventional and more aggressive approach towards full autonomy, such as the redefinition of the whole vehicle concept by taking the steering wheel out of the vehicles.

Though it is still not clear what the real intentions or business models of those IT/technology companies are, it is quite unlikely that they are aiming to expand their existing business by selling SDVs. A more likely scenario is to create new businesses by offering new online products and services related to SDVs, to reduce costs, or to become technology suppliers for the automotive industry. Due to their different business models, the SDVs might be designed to function within certain local or regional areas only, in order to reduce the system complexity and development/ testing effort. Such SDVs are most likely to be produced in small volumes and are therefore relatively less sensitive to local/regional specific customizations.

7.2.3 Transformative

The transformative strategy aims to realize full autonomy within a limited scope. By first tackling a simple scenario, such as by limiting the

Table 7.1   Brief comparison of SDV development strategies

Aspect	Evolutionary	Revolutionary	Transformative
Key Player	Auto industry	Non-automotive tech companies	High-tech start-ups
Operated by	Laypeople	Trained personnel and/or laypeople	Trained personnel
Geographical Range	Unlimited	Regional	Local

Source: Adapted from "Deployment Scenarios for Vehicles with Higher-Order Automation", by Sven Beiker, 2016, Springer Berlin Heidelberg, Berlin, Heidelberg, p. 193-211.

operation to small geographical areas and traveling at low speed, SDVs can be directly developed at high automation levels (level 4 or above) in a relatively short time. Support for more complex scenarios, e.g., larger operating areas, higher speed, mixed traffic, etc., can be gradually introduced as the SDV evolves over time.

The transformative strategy is typically pursued by high-tech startups, specializing in delivering SDV solution for a specific use-case, such as autonomous public shuttles and last mile delivery transports. Whereas SDV products developed by companies pursuing evolutionary and revolutionary approaches are targeted at laypeople, SDVs from the transformative approach typically require trained or skilled personnel for operation and monitoring.

Due to the high level of customization required for each local deployment, these start-ups are likely to position themselves as technology suppliers for service operators, such as local public transport authorities, logistic companies, amusement parks, etc.

Table 7.1 summarizes some key differences among the three SDV development strategies.

Which strategy will eventually bring level 5 SDVs faster to reality? Only time will tell.

7.3   Trends in deep learning for SDVs

Deep learning is undoubtedly one of the biggest technology buzzwords of recent years. This breakthrough technology has fueled significant improvements in modern applications, including face recognition [22], voice synthesis [23], and even brain tumor segmentation [12]. In the realm of SDV development, researchers and engineers around the world are embracing this technology to push SDV capabilities to the next level, especially in perception. One particular area of interest is feature learning, which makes object classification possible without humans having to manually define the features in advance. In other words, the computer learns the features on its own. As we already saw in Chapter 3, the quality of the feature engineering determines the performance of object detection. However, feature engineering is a largely manual process that requires deep knowledge and domain expertise. The impressive success of deep learning applications in recent years, especially in the area of image recognition, makes deep learning a promising technology to extend SDV capabilities beyond what is currently achievable using the conventional approach. Even though deep learning has the potential to become a key enabling technology in SDV, its deployment brings several challenges as well, which we will discuss in this section.

In the general context of Artificial Intelligence (AI), deep learning is a Machine-Learning (ML) method that utilizes Artificial Neural Networks (ANN) with a vast number of hidden layers, hence its synonym *deep neural networks*. Artificial intelligence is a sub-field of computer science that deals with enabling computers to carry out tasks in an intelligent way, usually by imitating human behavior and thinking processes [5]. One way to develop such AI systems is to employ ML algorithms that enable a computer to accomplish its goal by learning, i.e., without being explicitly programmed. This is a stark contrast to the 'manual' or rule-based approach to systems engineering, in which the computers' behaviors are based on rules defined and coded by human experts. The relationship between AI, ML, NNs, and deep learning is illustrated in Figure 7.9.

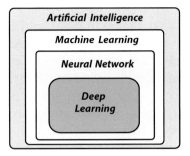

Figure 7.9: Relationship between deep learning, neural networks, machine learning and artificial intelligence. (Adapted from "Efficient Processing of Deep Neural Networks: A Tutorial and Survey", by Vivienne Sze, Yu-Hsin Chen, Tien-Ju Yang, Joel Emer, 2017, CoRR, abs/1703.09039. ©2017 Vivienne Sze, Yu-Hsin Chen, Tien-Ju Yang, Joel Emer)

Due to the many hidden layers and the sheer number of parameters involved, deep learning requires heavy computational resources, and a very large amount of training data. A deep neural network is made up of many types of layers. The *convolution* layer performs a series of mathematical operations on the input data. In many cases, the convolution is followed by some *activation* function that injects non-linearity into the model, and a *pooling* function that performs down-sampling of the input data for the sake of reducing the feature dimensions. The convolution, activation, and pooling layers are usually repeated many times, until they are fully connected in the last layer, where the classification is performed and the final result is provided.

Modeling and training deep neural networks is very resource intensive. In the early 2010s, it required 1000 CPUs worth upwards of one million USD to create an 'artificial brain' that could detect cat images quite impressively using deep learning [14]. Nowadays the use of cloud computing and hardware based on *Graphical Processing Units (GPUs)* for deep learning makes development faster and more affordable. Developers usually do not train their models from scratch, but instead take a pre-trained general model, and improve it to solve their specific problem. This approach is known as *transfer learning*.

There are several popular models or deep neural network architectures that have been proposed. One of the most popular is the *AlexNet* that significantly outperformed other competitors in the prestigious

ImageNet challenge in 2012 [18]. ImageNet is a large hand-annotated database of over 14 million images in more than 20 thousand categories, which has become the standard dataset for benchmarking deep learning models, and arguably the most significant innovation that sparked the deep learning revolution [9]. AlexNet comprises five convolutional layers and three fully connected layers, where the *Rectifier Linear Unit (ReLu)* activation function is applied after each convolutional layer. Another noteworthy architecture is *ResNet*, which uses residual networks to achieve excellent results using very deep models, but without suffering from the degradations that are usually observed in such very deep models [13]. AlexNet and ResNet architectures are depicted in Figure 7.10 and 7.11, respectively.

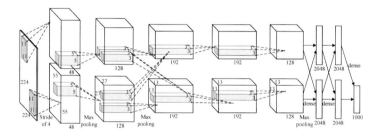

Figure 7.10: AlexNet architecture. (Adapted from "Optimized Compression for Implementing Convolutional Neural Networks on FPGA", by Min Zhang, Linpeng Li, Hai Wang, Yan Liu, Hongbo Qin, and Wei Zhao, 2019, Electronics 2019 8(3), p. 295. ©2019 Min Zhang, Linpeng Li, Hai Wang, Yan Liu, Hongbo Qin, and Wei Zhao, https://www.mdpi.com/electronics/electronics-08-00295/article_deploy/html/images/electronics-08-00295-g001.png, "Network architecture of AlexNet", https://creativecommons.org/licenses/by/4.0/legalcode)

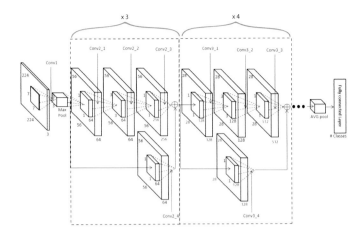

Figure 7.11: ResNet architecture with 50-layer residual nets (ResNet-50). (Adapted from "CNN-Based Multimodal Human Recognition in Surveillance Environments", by Ja Hyung Koo, Se Woon Cho, Na Rae Baek, Min Cheol Kim, and Kang Ryoung Park, 2018, Sensors 2018, 18(9), p. 3040. ©Ja Hyung Koo, SeWoon Cho, Na Rae Baek, Min Cheol Kim, and Kang Ryoung Park, https://www.mdpi.com/sensors/sensors-18-03040/article_deploy/html/images/sensors-18-03040-g008.png, "The structure of ResNet-50", Correction, https://creativecommons.org/licenses/by/4.0/legalcode)

One of the hot topics in deep learning is undoubtedly *Generative Adversarial Networks (GAN)* [11]. GANs, as shown in Figure 7.12, use two networks with opposing goals, the *generator* network and the *discriminator* network, which are trained to compete with each other until an equilibrium is reached. The generator's goal is to produce realistic samples from a vector of random noise to 'fool' the discriminator, while the discriminator aims to distinguish whether these inputs are fake or real, i.e., samples from a real dataset. Because the generator has significantly fewer parameters than the amount of data used for its training, the generator is forced to capture and 'compress' the essence of the data in order to generate samples correctly. Although the idea might sound simple, working with GANs is actually very challenging. GANs are hard to train, and reaching the equilibrium is not an easy task [20]. However, GANs have proved to produce impressive empirical results in *semi-supervised learning*, i.e., learning from mainly unlabeled data with some labeled data, and this makes it a promising technology that might push the capability of deep learning even further.

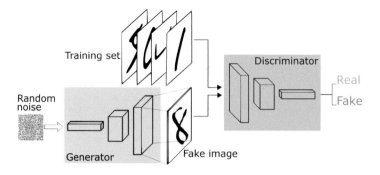

Figure 7.12: Generative adversarial networks. (Reprinted with permission from Thalles Silva. ©2017 Thalles Silva)

7.3.1 *Applying deep learning for SDVs*

In the context of applying deep learning to SDVs, there are two major paradigms observable within the SDV development community. The first paradigm is to apply deep learning to improve specific tasks within the SDV process chain, also known as *semantic abstraction learning*. The other paradigm is the *end-to-end learning*, i.e., taking raw sensor data as input, the vehicle control command as output and the whole process 'in-between' is learned using deep learning.

7.3.1.1 *Semantic abstraction learning*

Semantic abstraction learning refers to learning at a modular level or learning a specific task. The term module here does not necessarily mean a single component, it can also be a set of 'semantically meaningful' components that make up a specific functionality, such as object classification. Taking the object classification module as an example, the computer finds the optimal parameters for its deep neural network model by learning from the inputs and outputs of the module. A typical input in this case is raw image data or a 3D point cloud, depending on the sensor used. The output can be the class of detected object, e.g., pedestrian, car, etc., or class candidates with their corresponding confidence measures and perhaps their coordinates or bounding boxes. The rest of the processing components are beyond the scope of the learning, and they are not 'substituted' by deep learning.

7.3.1.2   End-to-end learning

The end-to-end paradigm learns the whole processing from input (raw sensor data) to output (vehicle control commands). In contrast to the semantic abstraction approach, the end-to-end paradigm aims to completely mimic the human decision-making process and driving behavior, based on how the environment is perceived by the sensors and issuing vehicle commands such as a human driver would in that particular situation. In other words, end-to-end learning is an alternative approach to solve the complex SDV problem in a holistic manner, without applying the standard engineering principle of problem decomposition [10].

7.3.2   Open questions

The success of deep learning, especially in image and speech recognition, has led researchers and engineers to intensively experiment with the technology to solve some of the hard problems in SDV development, or to improve the existing solutions. Deep learning has been prototyped to improve SDV perception, such as pedestrian detection [1], localization [17], or even to learn the whole data processing and decision making based on end-to-end learning [4].

Nevertheless, some important questions remain unanswered, for example: *to what extent can deep learning be safely deployed in safety-critical applications, such as SDVs?* In other words, how does the technology fit into current and future safety standards, e.g., ISO 26262, or national regulations? A functionality realized by deep learning is essentially a huge vector of parameters, which is completely incomprehensible to humans. This is quite the opposite to how safety-critical applications are engineered so far, which requires a thorough understanding of the function with its known limitations, potential hazards, etc.

Another open question is: *which of the two learning paradigms (semantic abstraction vs. end-to-end) will bring us to Level 5 Automation the soonest?* While the unconventional end-to-end approach might be necessary because the existing learning methods might not scale to arbitrarily complex tasks [10], critics of this approach argue that it might require significantly larger training datasets than its semantic abstraction counterpart, so that it becomes practically impossible to control the probability of system failure [21].

7.4 Summary

This chapter looked at some of the most promising use cases for SDV technology, both in the transportation context and in other contexts. Self-driving cars are the classic use case for SDVs; however, as we saw, their relative complexity may mean they are one of the last use cases to reach level 5 autonomy. Other transportation use cases like public shuttles and last-mile delivery vehicles seem more likely to achieve level 5 autonomy due to the limited scale of their operational environment. Beyond transportation, SDV technology promises to revolutionize agriculture and could be the basis for autonomous robots used for things like security patrolling. It will also have significant impact on the performance of autonomous rescue robots designed to replace or supplement the use of humans in hazardous situations.

We went on to compare and contrast the three approaches to SDV development, namely the Evolutionary, Revolutionary and Transformative approaches. These generally reflect the philosophies of traditional automobile manufacturers, multi-national technology firms, and start-ups, respectively. It is likely that the Transformative approach will be the quickest to achieve full automation; however, this is due to the small scale of the typical operating environment, along with the restricted use cases. By contrast, the Evolutionary approach will take a long time to achieve full automation, but the resulting SDV will be able to operate in any environment, and will be globally marketable, meeting all international safety standards.

Finally, we looked at how deep learning may revolutionize the pace of SDV development. Semantic abstraction promises to give excellent results for solving problems such as perception, while the end-to-end approach could lead to SDVs that perfectly mimic how a skilled human driver controls a car.

References

[1] Anelia Angelova, Alex Krizhevsky, Vincent Vanhoucke, Abhijit S Ogale, and Dave Ferguson. Real-time pedestrian detection with deep network cascades. In *BMVC*, volume 2, page 4, 2015.

[2] D. Baker and Scientific American. *Inventions from Outer Space: Everyday Uses for NASA Technology*. Universal International, 2000.

[3] Sven Beiker. *Deployment Scenarios for Vehicles with Higher-Order Automation*, pages 193–211. Springer Berlin Heidelberg, Berlin, Heidelberg, 2016.

[4] Mariusz Bojarski, Philip Yeres, Anna Choromanska, Krzysztof Choromanski, Bernhard Firner, Lawrence D. Jackel, and Urs Muller. Explaining how a deep neural network trained with end-to-end learning steers a car. *CoRR*, abs/1704.07911, 2017.

[5] Collins. Artificial intelligence definition and meaning. https://www.collinsdictionary.com/dictionary/english/artificial-intelligence. [accessed 24-Dec-2018].

[6] European Commission. Statistical pocketbook 2018 - EU transport in figures. https://ec.europa.eu/transport/sites/transport/files/pocketbook2018.pdf. [accessed 20-May-2018].

[7] DARPA. DARPA robotics challenge (DRC) (archived). https://www.darpa.mil/program/darpa-robotics-challenge. [accessed 20-May-2018].

[8] Arturo Davila, Eduardo del Pozo, Enric Aramburu, and Alex Freixas. Environmental benefits of vehicle platooning. Technical report, SAE Technical Paper, 2013.

[9] Jia Deng, Wei Dong, Richard Socher, Li-Jia Li, Kai Li, and Li Fei-Fei. Imagenet: A large-scale hierarchical image database. In *Computer Vision and Pattern Recognition, 2009. CVPR 2009. IEEE Conference on*, pages 248–255. IEEE, 2009.

[10] Tobias Glasmachers. Limits of end-to-end learning. *arXiv preprint arXiv:1704.08305*, 2017.

[11] Ian Goodfellow, Jean Pouget-Abadie, Mehdi Mirza, Bing Xu, David Warde-Farley, Sherjil Ozair, Aaron Courville, and Yoshua Bengio. Generative adversarial nets. In *Advances in neural information processing systems*, pages 2672–2680, 2014.

[12] Mohammad Havaei, Francis Dutil, Chris Pal, Hugo Larochelle, and Pierre-Marc Jodoin. A convolutional neural network approach to brain tumor segmentation. In *International Workshop on Brainlesion: Glioma, Multiple Sclerosis, Stroke and Traumatic Brain Injuries*, pages 195–208. Springer, 2015.

[13] Kaiming He, Xiangyu Zhang, Shaoqing Ren, and Jian Sun. Deep residual learning for image recognition. In *Proceedings of the IEEE conference on computer vision and pattern recognition*, pages 770–778, 2016.

[14] D Hernandez. Now you can build Google's $1 million artificial brain on the cheap. *Wired*, 6(3):413–421, 2013.

[15] ITF. Managing the transition to driverless road freight transport. https://www.oecd-ilibrary.org/content/paper/0f240722-en, 2017. [accessed 24-Dec-2018].

[16] Martin Joerss, Jürgen Schröder, Florian Neuhaus, Christopher Klink, and Florian Mann. Parcel delivery: The future of last mile. *McKinsey & Company*, 2016.

[17] Alex Kendall, Matthew Grimes, and Roberto Cipolla. Posenet: A convolutional network for real-time 6-dof camera relocalization. In *Proceedings of the IEEE international conference on computer vision*, pages 2938–2946, 2015.

[18] Alex Krizhevsky, Ilya Sutskever, and Geoffrey E Hinton. Imagenet classification with deep convolutional neural networks. In *Advances in neural information processing systems*, pages 1097–1105, 2012.

[19] Keiji Nagatani, Seiga Kiribayashi, Yoshito Okada, Kazuki Otake, Kazuya Yoshida, Satoshi Tadokoro, Takeshi Nishimura, Tomoaki Yoshida, Eiji Koyanagi, Mineo Fukushima, et al. Emergency response to the nuclear accident at the Fukushima Daiichi nuclear power plants using mobile rescue robots. *Journal of Field Robotics*, 30(1):44–63, 2013.

[20] Tim Salimans, Ian Goodfellow, Wojciech Zaremba, Vicki Cheung, Alec Radford, and Xi Chen. Improved techniques for training GANs. In *Advances in Neural Information Processing Systems*, pages 2234–2242, 2016.

[21] Shai Shalev-Shwartz and Amnon Shashua. On the sample complexity of end-to-end training vs. semantic abstraction training. *arXiv preprint arXiv:1604.06915*, 2016.

[22] Yaniv Taigman, Ming Yang, Marc'Aurelio Ranzato, and Lior Wolf. Deepface: Closing the gap to human-level performance in face verification. In *Proceedings of the IEEE conference on computer vision and pattern recognition*, pages 1701–1708, 2014.

[23] Aäron Van Den Oord, Sander Dieleman, Heiga Zen, Karen Simonyan, Oriol Vinyals, Alex Graves, Nal Kalchbrenner, Andrew W Senior, and Koray Kavukcuoglu. Wavenet: A generative model for raw audio. In *SSW*, page 125, 2016.

[24] John Walter. Help wanted: How farmers are tackling a labor shortage. `https://www.agriculture.com/farm-management/estate-planning/help-wanted-how-farmers-are-tackling-a-labor-shortage`. [accessed 24-Dec-2018].

[25] Alan FT Winfield, Marta Palau Franco, Bernd Brueggemann, Ayoze Castro, Miguel Cordero Limon, Gabriele Ferri, Fausto Ferreira, Xingkun Liu, Yvan Petillot, Juha Roning, et al. Eurathlon 2015: A multi-domain multi-robot grand challenge for search and rescue robots. In *Conference Towards Autonomous Robotic Systems*, pages 351–363. Springer, 2016.

Final words

I hope you have found this book useful. As mentioned in the beginning, this book is designed to bridge the gap between books and articles giving a general overview of self-driving vehicle technology and technical publications giving detailed knowledge on specific areas relating to SDVs.

The technology is still in its infancy, and many problems remain to be solved. But if you have the appropriate skills and creativity, I hope this book will encourage you to collaborate or to create side innovations based on the SDV technology. By doing so, you will be sharing in one of the most impactful innovations in personal mobility since the invention of the internal combustion engine.

Index